T0190095

Sustainable Textiles: Production, Processing, Manufacturing & Chemistry

Series Editor

Subramanian Senthilkannan Muthu, Head of Sustainability, SgT and API, Kowloon, Hong Kong

This series aims to address all issues related to sustainability through the lifecycles of textiles from manufacturing to consumer behavior through sustainable disposal. Potential topics include but are not limited to: Environmental Footprints of Textile manufacturing; Environmental Life Cycle Assessment of Textile production; Environmental impact models of Textiles and Clothing Supply Chain; Clothing Supply Chain Sustainability; Carbon, energy and water footprints of textile products and in the clothing manufacturing chain; Functional life and reusability of textile products; Biodegradable textile products and the assessment of biodegradability; Waste management in textile industry; Pollution abatement in textile sector; Recycled textile materials and the evaluation of recycling; Consumer behavior in Sustainable Textiles; Eco-design in Clothing & Apparels; Sustainable polymers & fibers in Textiles; Sustainable waste water treatments in Textile manufacturing; Sustainable Textile Chemicals in Textile manufacturing. Innovative fibres, processes, methods and technologies for Sustainable textiles; Development of sustainable, eco-friendly textile products and processes; Environmental standards for textile industry; Modelling of environmental impacts of textile products; Green Chemistry, clean technology and their applications to textiles and clothing sector; Eco-production of Apparels, Energy and Water Efficient textiles. Sustainable Smart textiles & polymers, Sustainable Nano fibers and Textiles; Sustainable Innovations in Textile Chemistry & Manufacturing; Circular Economy, Advances in Sustainable Textiles Manufacturing; Sustainable Luxury & Craftsmanship; Zero Waste Textiles.

More information about this series at http://www.springer.com/series/16490

Fieke Dhondt · Subramanian Senthilkannan Muthu

Hemp and Sustainability

Fieke Dhondt
Utrecht University
Utrecht, The Netherlands

Subramanian Senthilkannan Muthu
Head of Sustainability
SgT Group and API
Kowloon, Kowloon, Hong Kong

ISSN 2662-7108 ISSN 2662-7116 (electronic)
Sustainable Textiles: Production, Processing, Manufacturing & Chemistry
ISBN 978-981-16-3336-2 ISBN 978-981-16-3334-8 (eBook)
https://doi.org/10.1007/978-981-16-3334-8

This Springer imprint is published by the registered company Springer Nature Singapore Pte Ltd.
The registered company address is: 152 Beach Road, #21-01/04 Gateway East, Singapore 189721, Singapore

Contents

About the Authors

Fieke Dhondt graduated Cum Laude with a B.Sc. in Fashion and Textile Technologies at the Amsterdam Fashion Institute and continued studying for MSc Sustainable Business and Innovation at the Faculty of Geosciences at Utrecht University. Her bachelor's thesis was regarding the future of hemp fibres under changing climate conditions. Furthermore, she contributed to a new sustainability strategy for a multinational by conducting research in Bangladesh, China and Hong Kong.

Dr. Subramanian Senthilkannan Muthu holds a Ph.D. in Textiles Sustainability and has written around 100 books and 100 research publications. He is well known for his contributions in the field and has extensive academic and industrial experience. He is the Editor-in-Chief of the Textiles & Clothing Sustainability Journal.

Chapter 1
Introduction

1.1 History of Hemp

Industrial hemp (*Cannabis sativa* L.) is an annual grown plant which is harvested as a multipurpose crop [26]. The name cannabis found its origin from Indo-Germanic [27] or to be more precise from the Scythian folk [8].

Hemp probably originates from the temperate parts in Asia. From central Asia, it further spread to China, Indochina, Thailand, and the Malaysian regions [76]. Traces of hemp fruits have been discovered in 8260 BC at Okinoshima, central Japan [50]. The evidence of the use of hemp as a textile in ancient Chinese times ranges back to around 6000 years ago [76]. The crop was, later on, expanded to western Asia, Egypt, and Europe around 2000–1000 BC. Hemp textile remainings from the seventh-century BC were also found in grave mounds in Gordion Turkey [7]. There is even more evidence found of early uses of hemp textiles. A study on hemp textile in the time of the Vikings concluded that hemp was not only used for coarser products such as rows and sails but was also used for more delicate household textiles [62].

The introduction to North and South America came later, in 1545 and 1606 [76]. New England was the first place in North America where hemp was imported. Flax still took the overhand of fibre production in New England, but hemp popularity grew in the southern parts of North America. Hemp in South America first arrived in Chile and was brought by the Spanish [13]. An overview of hemp's history can be seen in Fig. 1.1.

The uses of hemp were not limited to textiles; the plant was also used as a Hebrew ritual for death [8] and medicinal use in ancient China [76].

Even though hemp has been used for several Millennia, the production and use of the fibre crop were the highest in the last three centuries [53], until the early 1900s. From then on, many countries banned the production and use of marijuana due to its psychoactive effects [35, 41, 54], these bans also outlawed and therefore harmed the development of hemp, under the pretext of marijuana [30].

A fabric deficit, due to cut off deliveries of fibres, lead to a renewal of hemp during the Second World War [64]. Shortly after the renewed interest, a decline in

F. Dhondt and S. S. Muthu, *Hemp and Sustainability*, Sustainable Textiles: Production, Processing, Manufacturing & Chemistry,
https://doi.org/10.1007/978-981-16-3334-8_1

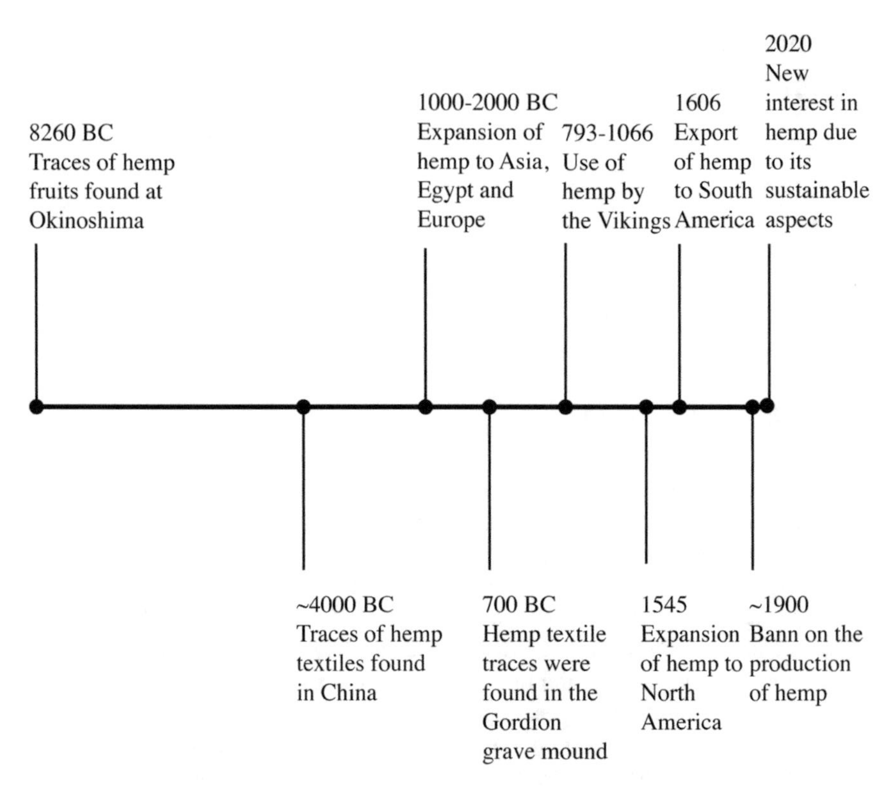

Fig. 1.1 History of hemp

production, caused by multiple factors, occurred again [53]. According to Ranalli and Ventur [53], the main reasons for this reoccurring decline were high labour cost, the introduction of man-made fibres, the association of the plant with illegal drugs, and the competition with cotton.

Harry Anslinger, Hearst, and Dupont had introduced the misunderstandings regarding hemp to limit the competition for Dupont's new patented products: plastics, paper from wood, and nylon. William Randolph Hearst, the owner of the Hearst newspaper, spread the inaccuracies in its articles, which were acknowledged as facts in the congressional testimony by Harry Anslinger. The widespread misunderstandings and unfounded arguments about cannabis hemp eventually led to the Marijuana Tax Act [30].

Recently, a renewed interest in hemp has grown, and this time it is due to the increased attention of sustainable fibres for textiles and the clothing sector. The sustainable properties of the hemp vis-a-vis other fibres caused this reacquired demand [26, 73]. Hemp has become attractive for the textile industry since it is a low-impact crop that barely needs any pesticides, herbicides, or fertilisers. Next to that, irrigation is hardly required due to the extensive root system of the crop [79]. These are some of the topmost properties of the crop in terms of sustainability metrics compared to other fibres used in the textiles and clothing sector.

1.2 General Properties of Hemp

Industrial hemp is classified as a C3 crop [79]. The first product that originates from carbon dioxide fixation defines if crops are C3 plants. C3 crops grow most efficient in cooler, more moist climates than at high temperatures [6].

The hemp plant consists out of five main parts: seeds, flowers, leaves, stem, and the roots [13]. The stem is layered as followed: the epidermis on the outer shell, cortex, bast fibres, wood core, and finally a hollow space [68]. The chemical composition of hemp is 67% cellulose, 16.1% hemicellulose, 0.8% pectins, 3.3% lignins, 2.1% water-soluble, 0.7% fat, and wax and 10% moisture according to the study of Turner [68]; these values can fluctuate depending on the hemp stain and climate conditions. Pectins and lignins are non-cellulosic components which can be considered as a kind of glue that keep the bast fibres and the wooden core of hemp together [68].

Industrial hemp is part of the Cannabaceae family [72] and contains the cannabinoids tetrahydrocannabinol (THC) and cannabidiol (CBD). THC is a psychoactive component of which hemp does not contain more than 0.3% [23], whereas the anti-psychoactive ingredient CBD has a higher presence in the crop [24]. Since the CBD content in hemp is much higher than the THC content, it is not possible to get intoxicated from this plant [33].

The best time to sow hemp crops in the temperate climate is around April, depending on the temperatures. However, planting past the first of May can result in lower dry yield and sowing to soon allows weeds to take the overhand [25]. In the first period of hemp growth, the crop grows relatively slow until 5–6 weeks after being sown [58]. Hemp plants grow the most in the last month and can double its length in that period. The length of the crop can reach up to 6 m, with an average growth of 10 cm per day [63]. Longer stems result in coarser fibres and are less suitable for textile end-use [68]. Fibre hemp is planted the closest together as possible for the highest yield [72]. A too high plant density will result in self-thinning of the stem [72] and lowers the plant weight, but only slightly impacts crop yield [4]. Hemp crops are sown at the end of August or the beginning of September [25]. It is best to harvest before flowering occurs since a later harvest will impact the stem part that is used for high-quality yarn production [75].

The carbon uptake of hemp can take up to 22 tonnes per hectare, which is considered to be higher than agroforestry [74]. Hemp is also considered sustainable due to its weed suppressing abilities [39], its soil-cleaning properties [36], its ability to benefit other plants it is grown in rotation with [65] and its positive impact on biodiversity [43]. Besides that, hemp can grow in many different climates [19].

1.2.1 Fibre Properties

Hemp textiles are differentiated from other fibres due to their features, particularly the aseptic characteristics, excellent absorbency and hygroscopicity, good thermal and electrostatic features, protect the wearer against UV radiation, and lack allergenic effects [34]. Some other qualities of hemp which have been highlighted in other literature are that hemp fibres are strong, keep their shape and hardly stretch. Hemp's high absorbency also makes the colour of dyed hemp textile last longer, and the colour stays more vibrant. Where hemp is often seen as a coarse and harsh textile, it is nowadays also possible to use it for softer and lighter fabrics. All these specific properties make hemp an interesting fibre for the textile industry [45].

1.2.2 Compared to Other Fibre Types in Terms of Sustainability Metrics

Hemp is seen as a sustainable alternative for natural and synthetic fibres [3]. Cotton and polyester are the currently most used fibres, whereas the bast fibre flax is a competitor of hemp [67]. Cotton is well known for its extensive water use [57], and use of insecticides and pesticides [55]. Polyester, on the other hand, is criticised due to microfibre shedding [56] and its link to the oil industry [57].

The most considerable difference between these four fibre types is between the synthetic fibre polyester and the three cellulosic fibres cotton, flax, and hemp. Whereas the cellulosic fibres grow as crop, polyester is made from crude oil. There is also a distinguishing between the cellulosic fibres, which is between the bast fibres and the seed fibre cotton. The fibres in the bast originate from the stem, and cotton fibres come from the seed boll of the cotton plant [18].

Yarn quality depends, among other things, on the type of fibres and can be measured according to various parameters of which some are shown in Table 1.1 [47]. As shown in Table 1.1, the diameter of hemp fibres has a relatively large span, but the diameter in microns of flax is even larger. Larger diameter spans are related to the coarseness of a fibre, the larger the diameter, the stiffer the fibre is [59]. Hemp fibres are overall longer than flax and cotton fibres. Long fibres show more cohesion than shorter fibres [44]. Next to that, the tenacity of hemp and flax is higher than the other two fibres. A higher tenacity results in a strong fibre but lowers the flexibility of the fibre [61]. The elongation at breakage is quite similar to the values of polyester but higher than the values of cotton and flax. This is mainly caused due to a larger difference of finesses of hemp fibres [66]. Moisture regains, also known as the absorbance rate, of hemp and flax are comparable. Moisture absorbency is an important factor for fabrics. Absorbency impacts the wearer comfort, static build-up, shrinkage, moisture repellence, and crease recovery of the textile [31].

Table 1.1 Values of fibre parameters of different textile fibres

	Hemp [66]	Flax	Cotton	Polyester
Diameter	15–50 microns	40–80 microns [11]	12–20 microns [11]	12–25 microns [11]
Length	150–250 cm	10–100 cm [11]	2.22–3.18 cm [11]	–
Tenacity	40–70 cN/Tex	Dry: 53–62 cN/Tex Wet 62–80 cN/Tex [20]	Dry: 26–43 cN/Tex Wet: 27–56 cN/Tex [2]	45 cN/Tex [12]
Elongation at break	23%	1.5–3.5% [20]	3–9.5% [2]	30% [12]
Moisture regains [78]	12%	12–14%	8.5%	0.4%

1.3 Hemp Textile Production

Hemp is an annual plant, sown in the spring and harvested in the fall [32]. As shown in Fig. 1.2, the most common textile processes used for hemp starts with cultivation, harvesting, retting, decortication [3, 32], hackling/drawing/carding, spinning, weaving/knitting, dyeing, and finishing [32, 45].

1.3.1 Harvesting

Hemp is harvested just before flowering, as the fibre quality decreases after flowering. Hemp can be harvested in various ways, depending on the end-use. The two most common harvesting techniques for hemp fibres are longitudinal and

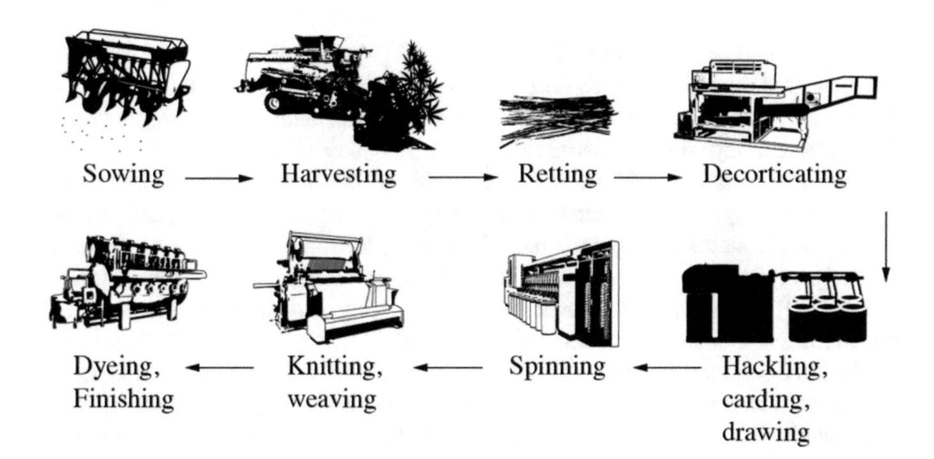

Fig. 1.2 Hemp production chain

disordered harvesting. The outcome of longitudinal harvesting is parallel fibre bundles, and the result from disordered harvesting is disarranged hemp stalks [3].

Longitudinal harvesting machines cut and bind the hemp stalks together. After this, the hemp bundles are left in the field for drying and later on portioned and formed into bunches or bales. Most of the machines can only handle stalks that are between 0.6 and 1 m long [3, 51], which is less efficient for hemp production, as hemp stalks can reach up to 6 m [63]. Longitudinal harvesting is mainly used for wet spun yarns, whereas disordered hemp harvesting can be performed by different machines. Harvesting can be done with a one knife cutting drum, harvesting system Bluecher 02/03 and multi-level cutter bars. The one knife cutting drum can be attached to tractors and cuts the stalks lengthwise and lays it in a windrow on the field under the tractor. The Bluecher 02/03 harvesting system cuts the stalk vertically on multiple places by guiding it through two cylinders before placing it on the land. The multiple level cutter can be connected to a tractor side attachment and cuts long hemp stalks in numerous parts [3].

Longitudinal harvesting results in longer fibres and higher quality, whereas cut stem harvest results in lower fibre quality. Nevertheless, longitudinal harvesting is less common, since most processing units cannot process long stalks. Cut stalk harvesting, on the other hand, is not often used for textile purposes due to low-quality fibres [3, 51].

Besides harvesting for fibre purpose, there are also machines which are specialised in harvesting both the seeds and the stalks. The seed heads are cut off first after that the machine cuts the stalks. The machine creates two windrows, one for seeds and one for the stalks [10]. The problem for dual-purpose hemp is often the difference in the maturity of seeds and stalks. Some of the dual-purpose machines solve this problem by cutting the stalks at a later stage than the seeds [3].

1.3.2 Retting

Retting is the process in which the pectins and lignins are dissolved, and the fibres are separated from the wooden core [32]. There are around eight different ways of hemp retting [26]. Dew retting and water retting are traditional retting methods; furthermore, enzyme retting is becoming more popular [77].

Dew retting is also referred to as field retting. The hemp stalks are laid in the field and are exposed to air after being harvested. Fungi and microorganism grow on the hemp stalks and release the fibres from the bast. This process depends strongly on local climate conditions, and only a few regions have the right climates [15, 29, 70]. Besides that, the length of the dew retting also influences the quality of the fibres and can take up to more than 50 days [37].

Another common method is water retting. This is a process where hemp stalks are retted in warm water tanks. Anaerobic bacteria grow in these tanks and thin the hemp stalks [16]. The study of Di Candilo et al. [14] analysed different retting times for water retting depending on the water type, temperature and bacteria. Stalks retted in pond water had the shortest retting time of 4 days with a temperature of 28 °C. The retting time of stems that were retted in well water was 12 days. Adding bacteria to well water retted stems at 20 °C, reduced the retting time to 6 days. Fibre quality can be better maintained with water retting then with dew retting, but high-labour cost and contamination of freshwater prevent this method from being widely used [5, 46, 70].

There are also options available that limit the waste of freshwater. One of the solutions could be the use of seawater instead of freshwater [80]. Another method to reduce costs and the use of non-renewable energy is to use thermal water [69], but this method would still contaminate freshwater.

Enzyme retting is becoming more popular than dew and water retting. Enzyme retting saves time, has higher quality fibres, and is eco-friendly [77]. This method is quite similar to water retting since the stalks are also sunk in a tank, but the tank is filled with commercial microbial enzymes [16]. Another enzyme retting method is spraying the stalks with enzymes till they are soaked and leave this for 2 min. It is crucial for enzyme retting to clean the stalks thoroughly after retting. The enzymes will otherwise continue operating in the stems [1]. A hydrothermal pre-treatment can be necessary for high-quality enzyme retted fibre [38].

1.3.3 Decorticator

Decorticating is the process in which the bast fibres and the woody core are separated. The wooden part is broken into multiple pieces, also called hurds, by break and kink forces [3, 32]. There are three different decorticating processes: breaking, milling, and scutching. In the breaking processes, stalks are guided through rollers that crush the stalks [70]. Another method, milling is more efficient than breaking and also cleans the stalks better. Lamentably, this method has the possibilities to damage fibres [3]. The last process, scutching, is more of a cleaning process that separates the remaining hurds and short fibres from the long fibres. In this process, fibre bundles are guided on rubber belts through rotating drums and beat with projecting bars [70]. Hemp fibres are more refined after this process [3]. The fibres are baled for storing after the decorticator has separated them [32].

1.3.4 Hackling, Carding, Drawing

The decorticated hemp is first cut and baled for three to ten days in high humidity circumstances before it can be hackled. Hackling is the processes in which the hemp fibres are combed. The hemp fibres are guided through metal pins which are attached to wooden boards. This process helps to separate the fibres and hurds even more and removes impurities. Long and short fibres are also further divided at this step. The end product after the hackling stage is hemp slivers [66].

In the carding process, fibres are once again cleaned, and all the fibre clusters become individual fibres. For this step, it is mainly important that all the fibres orientate in the same direction. This is achieved by a belt and cylinder with fine wires that separate and arrange the fibres into the right direction. The straight orientated fibres are turned into a twist less rope, also referred to as a carded sliver [21].

Drawing prepares slivers to a density level that is preferred for spinning purposes. Drawing involves two processes: doubling and drafting. Combining different slivers at the drawing processes is referred to as doubling. Drafting is the process in which the slivers are straightened and lengthened. Drafting happens once before hackling and twice after hackling and carding [21].

1.3.5 Spinning

The slivers are first going through the roving frame before they will be spun. Roving assures that the slivers are further lengthened and slightly twisted so they can be handled in the spinning process [21].

The most common spinning methods for hemp are wet and dry spinning [66]. During wet spinning, the roving passes through an immersion tank filled with substances that bond the fibres and makes them wet. Wet spinning reduces the number of breaks during spinning, which often happens for bast fibres since there is a higher unevenness of fineness [52]. Wet spinning has a negative environmental impact and therefore makes this a less interesting method for sustainable hemp yarn production [66]. The different ways of dry spinning for hemp can be ring or rotor spinning [71].

When ring spinning hemp fibres, the roving is fed into the drafting system in which the fibres are drawn and parallelised. The sliver is wound onto the tube using a rotating traveller on the rail. Through this process, the yarn receives its final twist. The ring spinning system is capable of producing very fine yarns. The ring spinning system can spin every fibre that is not from a tree. But there are also some downsides by ring spinning. The production speed in ring spinning is limited since the twisting and winding cannot be separated during ring spinning. Moreover, the possibilities for further automation are also too limited [21].

Engineers looked for an alternative for ring spinning despite all the good fibre properties from a ring-spun yarn, so they came up with the rotor spinner. The rotor spinning frame is fed with a card sliver or a draw frame sliver. The sliver is opened up to single fibres by an opening roller. This roller then cleans the fibres and separates the dirt. After this process, the fibres are transported to a centrifuging tube. The open end of the yarn is used to withdraw fibres while the rotor is twisting the thread when it is withdrawn. Spinning in which the yarn is assembled at such an open end is called open-end spinning systems. The strength of rotor spinning lies in its high productivity and its great flexibility regarding the raw material. The great advantage of rotor spinning is that the insertion of the twist is completely separated from the winding up of the spun yarn. This process makes the production speed much higher. The problem encountered in conventional ring spinning is that a certain percentage of fibres is not twisted in the yarn and thus does not contribute to its strength. The rotor spinning is ten times faster than the ring and compact spinning [21].

1.3.6 Knitting, Weaving, Dyeing, and Finishing

After the yarns are spun, they are ready for knitting or weaving depending on the various factors, including the end-use of the apparel product. Hemp is more often woven than knitted, but both options are possible. The fabric can be dyed after weaving or knitting, or the hemp can be yarn or fibre dyed before weaving or knitting. After knitting and weaving, the textile is ready for finishing.

The first stage of hemp fabric finishing is scouring. This diminishes the remaining dirt, pectins, and lignins. Another often used finishing for hemp is bleaching; this process is intended to produce a different shade level and to prepare hemp for dyeing. Hemp fabrics are dried after being sourced and bleached; this can cause a slight shrinkage of around 2.5% in the warp and 1.5% in the weft [66].

1.4 Geographical Areas of Hemp Growth

Hemp grows best in temperate climates but is not limited to growing in these areas, as shown in Fig. 1.3. There are around 47 countries where industrial hemp is cultivated for commercial or research purposes [22, 60]. The biggest hemp producers are China, South Korea, Russia, the USA, and Canada [45].

Around 25% of the global hemp market is cultivated in Europe, with France as the biggest producer. Most of the European hemp fibres are used for biocomposites, paper, and pulp, but textile end-use is less on the forefront. Hemp seeds and medicinal uses of hemp are gaining more interest in Europe [48]. Data of FAOSTAT [22] regarding hemp fibre and tow showed that one-third of the hectares used for European hemp production is for fibre and tow, and two-third is used for seed hemp. FAOSTAT includes hemp trade data of raw, retted, scutched, combed fibre, tow, and waste.

The largest cultivator of fibre hemp is China, which is growing around half of the world's supply. The USDA report on industrial hemp estimated that in 2019, 66,700 hectares—of which over 50% for fibre hemp is grown in China [42]. South Korea and Russia, other significant contributors to hemp production, harvested around 16,000 tonnes from 35,000 ha in 2019 [22]. This makes Asia one of the most prominent hemp producing continents. Other interesting countries in Asia for hemp growth are India and Nepal, as presented during the hemp summit (2019).

North and South America are upcoming hemp production countries. The 2018 hemp farming bill in the USA revived the interest in hemp cultivation. At this time,

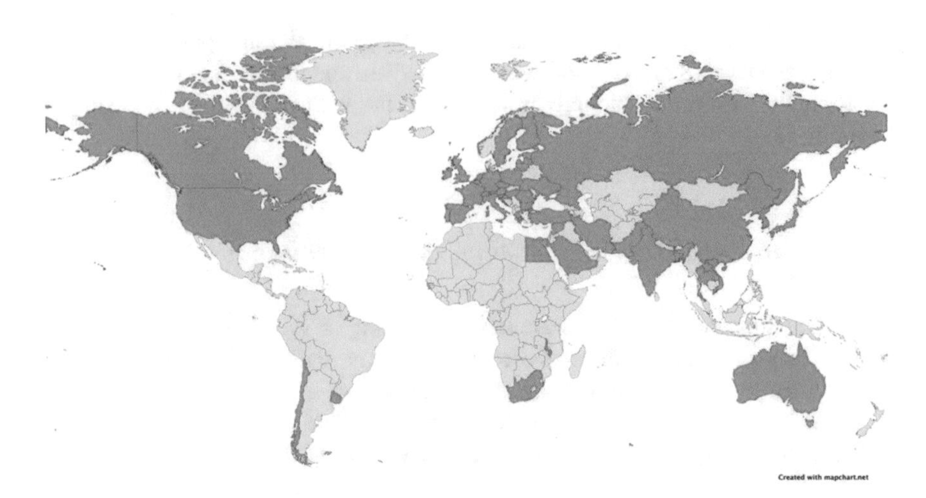

Fig. 1.3 Industrial hemp producing countries

most of the hemp farmers are interested in CBD hemp [9], but more growers are looking into alternative hemp products such as textiles. Canada already legalised growing hemp in 1998 and has a more established market [40]. Hemp was initially re-introduced in Canada for fibre purposes, but currently, only 22% of the cultivated hemp is for fibre use [28]. Central and South America are staying behind by North America, and hemp cultivation is less trending there. However, that can change in the future. Chile already allowed the growth of cannabis [22].

In Africa, industrial hemp production is still prohibited in various countries, which makes Africa the least involved continent in hemp fibre production [17, 22]. Hemp grew in South Africa for many centuries as a medicinal crop but has been prohibited from growing since 1928 [49]. The country recently recalled some parts of this ban and now allows the development, research, and cultivation of industrial hemp. 21 other African countries also requested for liberalisation of agricultural cannabis. Most of the countries are focusing on the pharmaceutical properties of the plant but in some other countries such as Malawi, and Uganda fibre hemp is also implemented. Famers are still obligated to request a licence before they can grow industrial hemp [17].

References

1. Akin DE et al (2004) Progress in enzyme-retting of flax. J Nat Fibers 1(1):21–47. https://doi.org/10.1300/J395v01n01_03
2. Allen HCJ (1958) Preparing cotton, web forming and bonding methods for cotton nonwovens. National Cotton Council, Raleigh. Available at: http://www.cotton.org/beltwide/proceedings/getPDF.cfm?year=1998&paper=P020.pdf. Accessed 9 Nov 2020

3. Amaducci S, Gusovius H (2010) Hemp—cultivation, extraction and processing. In: Müssig J (ed) Industrial applications of natural fibres. Wiley, Bremen, pp 109–134
4. Amaducci S, Errani M, Venturi G (2002) Response of hemp to plant population and nitrogen fertilization. Ital J Agron 6(2):103–111. Available at: https://www.academia.edu/19208994/Response_of_hemp_to_plant_population_and_nitrogen_fertilisation. Accessed 4 Nov 2020
5. Bacci L et al (2010) Effect of different extraction methods on fiber quality of nettle (*Urtica dioica* L.). Text Res J 81(8):827–837. https://doi.org/10.1177/0040517510391698
6. Bear R et al (2016) Principles of biology. New Prairie Press, Manhattan
7. Bellinger L (1962) Textiles from Gordion. Bull Needle Bobbin Club 46:4–33
8. Benet S (1975) Early diffusion and folk uses of hemp. In: Rubin V, Comitas L (eds) Cannabis and culture. De Gruyter Mouton, Berlin, pp 39–49
9. Brightfield Group (2019) U.S. hemp cultivation landscape. Brightfield Group, Chigaco. Available at: https://global-uploads.webflow.com/596691afde3c5856d866ae50/5d936f60b77add18c7a6a9e8_Hemp%20Cultivation%20Free%20Report%20(2).pdf. Accessed 05 Nov 2020
10. Chen Y, Liu J (2003) Development of a windrower for dual-purpose hemp (*Cannabis sativa*). Can Biosyst Eng (Le génie des biosystèmes au Canada) 45:2.1–2.7
11. Dai X-Q (2006) Fibers. In: Li Y, Dai X (eds) Biomechanical engineering of textiles and clothing. Woodhead Publishing, Cambridge, pp 163–177
12. Debnatha S, Sengupta S (2009) Effect of linear density, twist and blend proportion on some physical properties of jute and hollow polyester blended yarn. Indian J Fibre Text Res 34(1): 11–19
13. Dewey LH (1913) Hemp. In: Yearbook of the United States Department of Agriculture. Government Printing Office, Washington, D.C., pp 283–346
14. Di Candilo M et al (2000) Preliminary results of tests facing with the controlled retting of hemp. Ind Crops Prod 11(2–3):197–203. https://doi.org/10.1016/S0926-6690(99)00047-3
15. Donaghy JA, Boomer JH, Haylock RW (1992) An assessment of the quality and yield of flax fiber produced by the use of pure bacterial cultures in flax rets. Enzyme Microb Technol 14(2):131–134. https://doi.org/10.1016/0141-0229(92)90170-S
16. Donaghy JA, Levett PN, Haylock RW (1990) Changes in microbial populations during anaerobic flax retting. J Appl Bacteriol 69(5):634–641. https://doi.org/10.1111/j.1365-2672.1990.tb01556.x
17. Duvall CS (2019) A brief agricultural history of cannabis in Africa, from prehistory to canna-colony. EchoGéo, vol 48. Available at: http://journals.openedition.org/echogeo/17599. Accessed 10 Nov 2020
18. Eberle H et al (2014) Clothing technology—from fibre to fashion, 6th edn. Verlag Europa-Lehrmittel, Haan-Gruiten
19. Ehrensing DT (1998) Feasibility of industrial hemp production in the United states pacific north west. Oregon State University, Oregon. Available at: https://www.ers.usda.gov/publications/pub-details/?pubid=41757. Accessed 5 Nov 2020
20. El Mogahz YE (2009) Structure, characteristics and types of fiber for textile product design. In: El Mogahz YE (ed) Engineering textiles. Woodhead Publishing, Cambridge, pp 208–239
21. Elhawary IA (2015) Fibre to yarn. In: Sinclair R (ed) Textiles and fashion. Woodhead Publishing, Cambridge, pp 191–212
22. FAOSTAT (2020) Crops hemp tow waste. Available at: http://www.fao.org/faostat/en/#data/QC. Accessed 05 Nov 2020
23. Fournier G et al (1987) Identification of a new chemotype in cannabis sativa: cannabigerol—dominant plants, biogenetic and agronomic prospects. Planta Med 53(03):277–280. https://doi.org/10.1055/s-2006-962705
24. Fournier G (1981) Les chimiotypes du chanvre (*Cannabis sativa* L.) Intérêt pour un programme de sélection. Agronomie EDP Sci 1(8):679–688. https://doi.org/10.1051/agro:19810809
25. Friederich JC (1964) Enkele ervaringen met de teelt van hennep. Landbouwvoorlichting, vol 21, pp 145–149. Available at: https://edepot.wur.nl/368765. Accessed 2 Nov 2020

26. Gedik G, Avinc O (2020) Hemp fiber as a sustainable raw material source for textile industry: can we use its potential for more eco-friendly production? In: Muthu SS, Gardetti MA (eds) Sustainability in the textile and apparel industries. Springer, Cham, pp 87–109
27. Godwin H (1967) The ancient cultivation of hemp. Antiquity 41(161):42–49. https://doi.org/10.1017/S0003598X00038928
28. Health Canada (2020) Industrial hemp licensing statistics. Available at: https://www.canada.ca/en/health-canada/services/drugs-medication/cannabis/producing-selling-hemp/about-hemp-canada-hemp-industry/statistics-reports-fact-sheets-hemp.html. Accessed 5 Nov 2020
29. Henriksson G et al (1997) Identification and retting efficiencies of fungi isolated from dew-retted flax in the United States and Europe. Appl Environ Microbiol 63(10):3950–3956. https://doi.org/10.1128/AEM.63.10.3950-3956.1997
30. Herer J (2006) The emperor wears no clothes, 11th edn. AH HA Publishing, Austin
31. Hu JY, Li YI, Yeung KW (2006) Liquid moisture transfer. In: Li Y, Wong A (eds) Clothing biosensory engineering. Woodhead Publishing, Cambridge, pp 218–234
32. Jenkins T, Calfee L (2019) Hemp production review of literature with specified scope. Fibershed, San Geronimo. Available at: http://fibershed.org/wp-content/uploads/2019/01/hemp-literature-review-Jan2019.pdf. Accessed 31 Oct 2020
33. Karniol IG et al (1975) Effects of Δ9-tetrahydrocannabinol and cannabinol in man. Pharmacology 13(6):502–512. https://doi.org/10.1159/000136944
34. Kostic M, Pejic B, Skundric P (2008) Quality of chemically modified hemp fibers. Biores Technol 99(1):94–99. https://doi.org/10.1016/j.biortech.2006.11.050
35. Lee MA (2012) Smoke signals: a social history of marijuana—medical, recreational and scientific. Scribner, New York
36. Linger P, Müssig J, Fischer H, Kobert J (2002) Industrial hemp (*Cannabis sativa* L.) growing on heavy metal contaminated soil: fibre quality and phytoremediation potential. Ind Crops Prod 33(1):33–42. https://doi.org/10.1016/S0926-6690(02)00005-5
37. Liu M et al (2015) Effect of harvest time and field retting duration on the chemical composition, morphology and mechanical properties of hemp fibers. Ind Crops Prod 69:29–39. https://doi.org/10.1016/j.indcrop.2015.02.010
38. Liu M et al (2016) Controlled retting of hemp fibres: effect of hydrothermal pre-treatment and enzymatic retting on the mechanical properties of unidirectional hemp/epoxy composites. Compos A Appl Sci Manuf 88:253–262. https://doi.org/10.1016/j.compositesa.2016.06.003
39. Lotz LAP, Groeneveld RMW, Habekotte B, Oene H (1991) Reduction of growth and reproduction of cyperus esculentus by specific crops. Weed Res 31(3):153–160. https://doi.org/10.1111/j.1365-3180.1991.tb01754.x
40. Lupescu M (2019) Industrial hemp production trade and regulation. USDA, Ottawa. Available at: https://apps.fas.usda.gov/newgainapi/api/report/downloadreportbyfilename?filename=Industrial%20Hemp%20Production%20Trade%20and%20Regulation_Ottawa_Canada_8-26-2019.pdf. Accessed 5 Nov 2020
41. Manning P (2013) Drugs and popular culture: drugs, media and identity in contemporary society, 2nd edn. Routledge, New York
42. Mcgrath C (2020) 2019 hemp annual report—Peoples Republic of China. USDA, Washington D.C. Available at: https://apps.fas.usda.gov/newgainapi/api/Report/DownloadReportByFileName?fileName=2019%20Hemp%20Annual%20Report_Beijing_China%20-%20Peoples%20Republic%20of_02-21-2020. Accessed 7 Nov 2020
43. Montford S, Small E (1999) A comparison of the biodiversity friendliness of crops with special reference to hemp (*Cannabis sativa* L.). J Int Hemp Assoc 6(2):53–63. Available at: http://www.internationalhempassociation.org/jiha/jiha6206.html. Accessed 31 Oct 2020
44. Morton WE, Hearle JWS (2008) Physical properties of textile fibres, 4th edn. Woodhead Publishing, Cambridge
45. Muzyczek M (2020) The use of flax and hemp for textile applications. In: Kozlowski RM, Mackiewicz-Talarczyk M (eds) Handbook of natural fibres, vol 2: processing and applications. Taylor & Francis Group, Poznan, pp 147–168

46. Mwaikambo LY (2006) Review of the history, properties and application of plant fibres. Afr J Sci Technol 7(2):120–133. Available at: https://www.researchgate.net/publication/284760719_Review_of_the_history_properties_and_application_of_plant_fibres. Accessed 3 Nov 2020
47. Neckář B, Vyšanská M (2012) Simulation of fibrous structures and yarns. In: Veit D (ed) Simulation in textile technology. Woodhead Publishing, Cambridge, pp 222–265
48. New Frontier Data (2019) Hemp cultivation in Europe. Available at: https://newfrontierdata.com/cannabis-insights/developed-global-markets-hemp-acreage-comparison/. Accessed 05 Nov 2020
49. Ngobeni ND, Mokoena ML, Funnah SM (2016). Growth and yield response of fibre hemp cultivars (*Cannabis sativa* L.) under different N-levels in Eastern Cape Province of South Africa. Afr J Agr Res 11(2):57–64. https://doi.org/10.5897/AJAR12.0675
50. Okazaki H et al (2011) Early Holocene coastal environment change inferred from deposits at Okinoshima archeological site, Boso Peninsula, central Japan. Quatern Int 230:87–94. https://doi.org/10.1016/j.quaint.2009.11.002
51. Pari L, Baraniecki P, Kaniewski R, Scarfone A (2015) Harvesting strategies of bast fiber crops in Europe and in China. Ind Crops Prod 68(1):90–96. https://doi.org/10.1016/j.indcrop.2014.09.010
52. Racu C, Diaconescu R, Grigoriu A-M, Grigoriu A (2010) Optimization of hemp yarn grafting degree for medical textiles during simultaneous wet spinning-grafting. Cellul Chem Technol 44(9):365–368. Available at: https://www.cellulosechemtechnol.ro/pdf/CCT9(2010)/p365-368.pdf. Accessed 6 Nov 2020
53. Ranalli P, Venturi G (2004) Hemp as a raw material for industrial applications. Euphytica 140:1–6. https://doi.org/10.1007/s10681-004-4749-8
54. Rawson JM (2005) Hemp as an agricultural commodity. Congressional Research Service The Library of Congress, Washington, D.C. Available at: https://fas.org/sgp/crs/misc/RL32725.pdf. Accessed 7 Nov 2020
55. Rex D et al (2013) State of the art report. Mistra Future Fashion, Stockholm. Available at: http://mistrafuturefashion.com/wp-content/uploads/2015/12/D2.1-D4.1-Joint-state-of-the-art-report-MiFuFa-P2-P4-1.pdf. Accessed 11 Nov 2020
56. Roos S, Arturin OL, Hanning A (2017) Microplastics shedding from polyester fabrics. Mistra Future Fashion, Stockholm. Available at: http://mistrafuturefashion.com/wp-content/uploads/2017/06/MFF-Report-Microplastics.pdf. Accessed 12 Nov 2020
57. Sandin G, Roos S, Johansson M (2019) Environmental impact of textile fibers – what we know and what we don't know. Mistra Future Fashion, Stockholm. Available at: http://mistrafuturefashion.com/wp-content/uploads/2019/03/Sandin-D2.12.1-Fiber-Bibel-Part-2_Mistra-Future-Fashion-Report-2019.03.pdf. Accessed 12 Nov 2020
58. Sankari HS, Mela TJN (1998) Plant development and stem yield of non-domestic fibre hemp (*Cannabis sativa* L.) cultivars in long-day growth conditions in finland. J Agron Crop Sci 181:153–159. https://doi.org/10.1111/j.1439-037X.1998.tb00411.x
59. Saville BP (1999) Physcial testing of textiles. CRC Press & Woodhead Publishing, Boca Raton
60. Schluttenhofer C, Yuan L (2017) Challenges towards revitalizing hemp: a multifaceted crop. Trends Plant Sci 22(11):917–929. https://doi.org/10.1016/j.tplants.2017.08.004
61. Shuvo II (2020) Fibre attributes and mapping the cultivar influence of different industrial cellulosic crops (cotton, hemp, flax, and canola) on textile properties. Bioresour Bioprocess 7(51). https://doi.org/10.1186/s40643-020-00339-1
62. Skoglund G, Nockert M, Holst B (2013) Viking and early middle ages Northern Scandinavian textiles proven to be made with hemp. Sci Rep 3(1):1–6. https://doi.org/10.1038/srep02686
63. Small E, Catling PM (2008) Blossoming treasures of biodiversity: 27. Cannabis—Dr. Jekyll and Mr. Hyde. Biodiversity 31–38(1):31–38. https://doi.org/10.1080/14888386.2009.9712635
64. Small E, Marcus D (2002) Hemp: a new crop with new uses for North America. In: Janick J, Whipkey A (eds) Trends in new crops and new uses. ASHS Press, Alexandria, pp 284–326

65. Smith-Heisters S (2008) Environmental costs of hemp prohibition in the United States. J Ind Hemp 13(2):157–170. https://doi.org/10.1080/15377880802391308
66. Sponner J, Toth L, Cziger S, Franck RR (2005) Hemp. In: Franck RE (ed) Bast and other plant fibres. Woodhead Publishing Series in Textiles, Cambridge, pp 176–206
67. Textile Exchange (2020) Preferred fiber & materials market report 2020. Textile Exchange, Lamesa. Available at: https://store.textileexchange.org/product/2020-preferred-fiber-materials-report/. Accessed 2 Nov 2020
68. Turner AJ (1949) The structure of textile fibres. VIII–the long vegetable fibres. J Text Inst Proc 40(10):972–984. https://doi.org/10.1080/19447014908664732
69. Turunen L, van der Werf HMG (2007) The production chain of hemp and flax textile yarn and its environmental impacts. J Ind Hemp 12(2):43–66. https://doi.org/10.1300/J237v12n02_04
70. USDA (2000) Industrial hemp in the United States. USDA, Washington, D.C. Available at: https://www.ers.usda.gov/webdocs/publications/41740/15867_ages001e_1_.pdf?v=4861.7. Accessed 7 Nov 2020
71. Van Dam J, Van den Oever M, Oldenburger E, Reinders M (2018) Hemp for sustainable textile market developments. In: 15th international conference EIHA. European Industrial Hemp Association, Köln. Available at: http://eiha.org/media/2018/07/Jan_van_Dam-WUR-EIHA_2018.pdf. Accessed 6 Nov 2020
72. Van der Werf HMG (1994) Crop physiology of fibre hemp (*Cannabis sativa* L.). Wageningen Agricultural University, Wageningen. Available at: https://edepot.wur.nl/202103. Accessed 3 Nov 2020
73. Van der Werf HMG (2004) Life cycle analysis of field production of fibre hemp, the effect of production practices on environmental impacts. Euphytica 140(1):13–23. https://doi.org/10.1007/s10681-004-4750-2
74. Vosper J (2011) The role of industrial hemp in carbon farming. GoodEarth Resources PTY Ltd., Sydney. Available at: https://hemp-copenhagen.com/images/Hemp-cph-Carbon-sink.pdf. Accessed 31 Oct 2020
75. Westerhuis W (2016) Hemp for textiles: plant size matters. Wageningen University, Wageningen. Available at: https://edepot.wur.nl/378698. Accessed 6 Nov 2020
76. Wulijarni-Soetjipto N, Subarnas A, Horsten S, Stutterheim N (1999) *Cannabis sativa* L. In: de Padua L, Bunyapraphatsara N, Lemmens R (eds) Plant resources of South-East Asia: No 12 (1) medicinal and poisonous plants 1. Backhuys Publishers, Leiden, pp 167–175
77. Yadav D et al (2016) Potential of microbial enzymes in retting of natural fibers: a review. Curr Biochem Eng 3(2):89–99. https://doi.org/10.2174/221271190302160607151925
78. Yuying Z, Ruzhen D (2008) Conventional moisture regains of textiles. General Administration of Quality Supervision, Inspection and Quarantaine of the People's Republic of China, Beijing
79. Zatta A, Monti A, Venturi G (2012) Eighty years of studies on industrial hemp in the Po Valley (1930–2010). J Nat Fibers 9(3):180–196. https://doi.org/10.1080/15440478.2012.706439
80. Zhang LL et al (2008) Seawater-retting treatment of hemp and characterization of bacterial strains involved in the retting process. Process Biochem 43(11):1195–1201. https://doi.org/10.1016/j.procbio.2008.06.019

Chapter 2
The Environmental and Social Impacts of Hemp

2.1 Life Cycle Assessments

Life cycle assessment (LCA) is a scientific tool to evaluate the environmental performance of a product in its entire life cycle, from cradle to grave stages or cradle to gate. The assessment includes phases such as raw material extraction, material processing, product manufacture, consumer use, and end-of-life. The assessment looks at the inputs and outputs of the entire system for all life cycle phases such as raw materials, chemicals, accessories, all types of wastes, emissions to air, water and soil, as well as energy and energy use. According to the ISO standard 14040:2006 [22], LCA consists of four phases:

1. Goal and scope definition—The first step of the LCA, which describes the principles and framework. This step includes the goal and scope, the relationship between different LCA phases, applications, and limitations.
2. Life cycle inventory analysis—The second phase of the LCA includes the input and output data of the studied life cycle system. This phase collects the relevant environmental aspects of the studied system.
3. Life cycle impact assessment—The third phase is the life cycle impact assessment. The life cycle assessment provides information to assess the life cycle inventory results. In this way, the environmental impact of the outcome can be better explained. Some methods for the life cycle impact assessment are as follows:

    - Greenhouse gas protocol—The greenhouse gas protocol method quantifies the amount of CO_2 equivalent and divides the carbon dioxide emissions into fossil, biogenic, land change, and uptake. It also reflects the impact of a product on climate change [47].
    - Cumulative energy demand (CED)—The method of cumulative energy demand indicates the energy needs of the product life cycle. The method

F. Dhondt and S. S. Muthu, *Hemp and Sustainability*, Sustainable Textiles: Production, Processing, Manufacturing & Chemistry,
https://doi.org/10.1007/978-981-16-3334-8_2

divides the direct and indirect energy sources. The CED also separates renewable and non-renewable energy sources [43, 47].

- Eco indicator 99—Another method is the eco indicator 99. This method groups and weights all the included impact categories. Eco indicator 99 is a damages assessment and considers the damage to human health, ecosystem quality, and resources according to the impact categories. The outcome of this method is a single score for the entire life cycle of the product [1, 47].

4. Life cycle interpretation—The final phase of the LCA is the life cycle interpretation. The life cycle inventory analysis and the life cycle assessment results are summarised and discussed in this phase. After this phase, conclusions, recommendation, and a decision can be made.

The four phases review the environmental impact of the current production methods and uncover the present environmental hot spots. In this way, those involved in the product life cycle can improve current production methods. Implementation of enhanced manufacturing can eventually lower specific production methods' environmental impact [1]. Each type of life cycle assessment considers various impact indicators or impact categories for its study, depending on the goal and scope of the study. The following impact indicators or impact categories are considered in the life cycle assessments of hemp:

- Global warming potential—Global warming potential is the quantity or the number of greenhouse gases emitted that impact the atmosphere's heat radiation absorption. The unit is kg of CO_2 equivalent over 20, 100, or 500 years. Most of the LCA papers studied in this chapter used CO_2 equivalent over 100 years [1, 40–43]. Some studies also referred to this category as climate change [40–42].
- Terrestrial acidification—Acidifying pollutants, such as nitrogen and sulphur oxides impact the soil, organisms, and ecosystems. This category is measured in kg SO_2 equivalent [1, 17, 40].
- Eutrophication—Freshwater eutrophication and marine eutrophication are the two types of eutrophication considered in the LCAs of hemp fibre and textile production. Eutrophication in freshwater is the impact of the macronutrient's nitrogen and phosphorus in the environment. An increased amount of these macronutrients can cause a growth of biomass formation and disturb species' natural balance. The unit of this category is kg PO_4 equivalent [1, 17, 40]. Marine eutrophication is saltwater that enriches with nitrogen, phosphorus, and other plant nutrients. These nutrients stimulate the growth, primary production and blooms of algae, organism balance changes, and water quality degradation [44]. Marine eutrophication is measured in kg N equivalent [43].
- Abiotic depletion—Abiotic depletion depletes abiotic resources such as minerals, clay, and peat. The measurement unit is kg of antimony equivalent [1].
- Ozone depletion—Ozone molecules deplete when they come into contact with chlorine and bromine atoms. The ozone layer can reduce faster than ozone molecules create. The unit is kg of CFC-11 equivalent [1].

- Photochemical oxidation—The combination of volatile organics and nitrogen oxides in combination with sunlight creates photo-oxidants. Photo-oxidant formation produces reactive chemical compounds through ultraviolet light that negatively influence human health, ecosystems, and crops. The unit is kg ethylene (C_2H_4) equivalent [1, 17].
- Ecotoxicity—Ecotoxicity is the impact of toxic substances on specific environments. This category subdivides into terrestrial ecotoxicity, freshwater ecotoxicity, and marine aquatic ecotoxicity. The unit of ecotoxicity is kg of 1,4-dichlorobenzene equivalent [1, 42].
- Human toxicity—Human toxicity is the effect of toxic emission on the health of humans. The unit of human toxicity is kg of 1,4-dichlorobenzene equivalent [1].
- Particulate matter formation—Particulate matters are solid particles or liquid droplets. Particulate matters in the atmosphere are aerosols. The general diameter of particulate matter is ten μm or less. Natural particulate matter is volcanic ash, pollen, sea salt, sand, and others. Anthropogenic sources are soot, smog, chemical mist, and others [32]. The unit of particulate matter formation is kg PM10 equivalent [43].
- Energy use—The use of energy is the depletion of energetic resources [40]. Some studies also refer specifically to non-renewable energy use. Non-renewable energy is the use of coal, crude oil, natural gas, or uranium [17]. The unit of energy use is either MJ or GJ [17, 40].
- Land use—Land use or occupation is the use of land as a resource. Due to this, the land is unavailable for other practices during the crop growth period. The unit of land use is m^2 of occupied land per year [40].

Some of these categories are interrelated. The impacts of energy use, global warming potential, and acidification are corresponding. A large part of energy use comes from fossil fuels, and the burning of fossil fuels emits carbon dioxide and sulphur oxides. Carbon dioxide is a greenhouse gas, whereas sulphur oxides contribute to acidification [41].

Current hemp life cycle assessment studies are only conducted in Asia and Europe. There are currently no life cycle assessments of hemp production in North America, South America, or Africa. Therefore, the two continents that are reviewed are Asia and Europe.

2.2 Life Cycle Assessment Studies of Hemp Conducted in Asia

Asia is the most prominent hemp fibre producer [14]. China is the largest cultivator and produces more than half of the total fibre hemp quantities [27]. The product life cycle stages of the LCA of hemp grown in China [43] are displayed in Fig. 2.1.

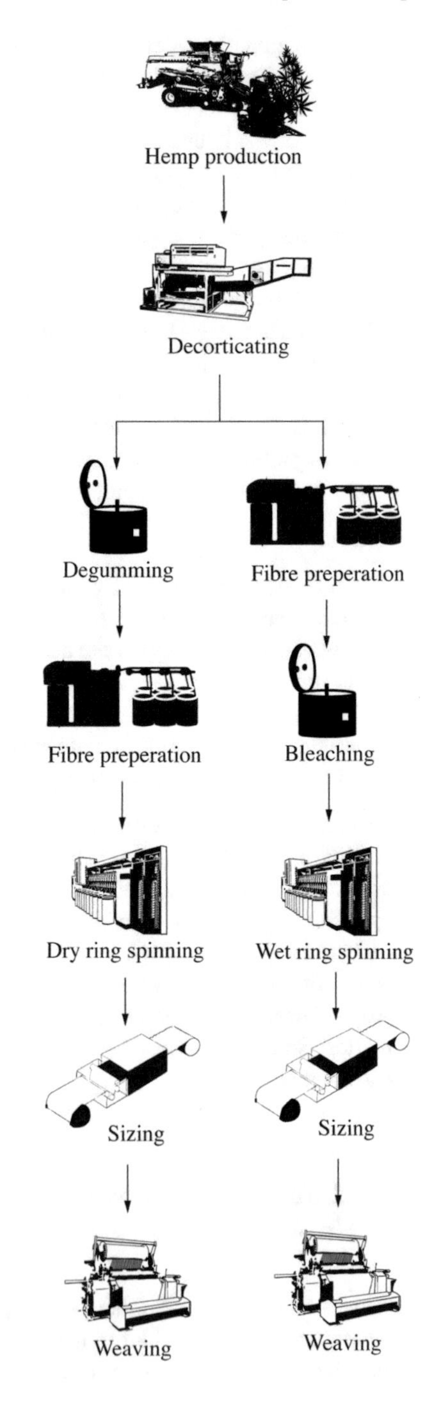

Fig. 2.1 Product life cycle stages of hemp produced in China

2.2.1 China

A LCA conducted by Van Eynde [43] studied the environmental impact of hemp textiles produced in China compared to cotton textiles. Two types of hemp textile were assessed: degummed and bleached hemp. The study included the global warming potential, terrestrial acidification potential, freshwater eutrophication potential, marine eutrophication potential, human toxicity potential, particulate matter formation potential, freshwater ecotoxicity potential, marine ecotoxicity potential, and fossil depletion potential as impact assessment indicators in the study.

A remarkable aspect of this study is that the production of degummed hemp has a lower impact on almost all measured categories compared to cotton and bleached hemp. Degummed hemp only had a larger impact than cotton on freshwater eutrophication potential, freshwater ecotoxicity potential, marine ecotoxicity potential, and fossil depletion potential. The freshwater eutrophication, freshwater ecotoxicity, and marine ecotoxicity were higher for degummed hemp than for cotton due to the production step degumming.

Bleached hemp had the highest environmental impact on eight out of nine impact categories. Bleached hemp had a higher impact than degummed hemp and cotton due to the bleaching stage, which has a large impact on the environment. Furthermore, the weaving and sizing stage for bleached hemp also had a large impact compared to the same stage for degummed hemp and cotton. Cotton only scored higher for marine eutrophication potential due to the fibre processing stage of cotton, which involves high fertiliser emissions.

For the processes after cultivation, hemp scored most of the times a worse or similar environmental performance as cotton. Especially the production step, degumming and bleaching, which was not included for cotton but only applicable for hemp, increased hemp's environmental impacts. There was also a difference in the results between degummed or bleached hemp. As degummed hemp overall scored a better performance than bleached hemp and in five out of nine categories also better than cotton. The study of Van Eynde [43] concluded that hemp has a lower impact regarding cultivation compared to cotton. The results were 50–90% lower than the environmental impact of cotton cultivation. But the old technology that is used for hemp fibre processing diminished hemp's environmental performance, which caused hemp to have worse environmental performance than cotton in some categories.

2.3 Life Cycle Assessments of Hemp Conducted in Europe

This section includes several life cycle assessments that were conducted according to European hemp production practices. The life cycle assessments are categorised according to the country they were assessed in and sorted according to the production processes they include, which are displayed in Fig. 2.2. The life cycle

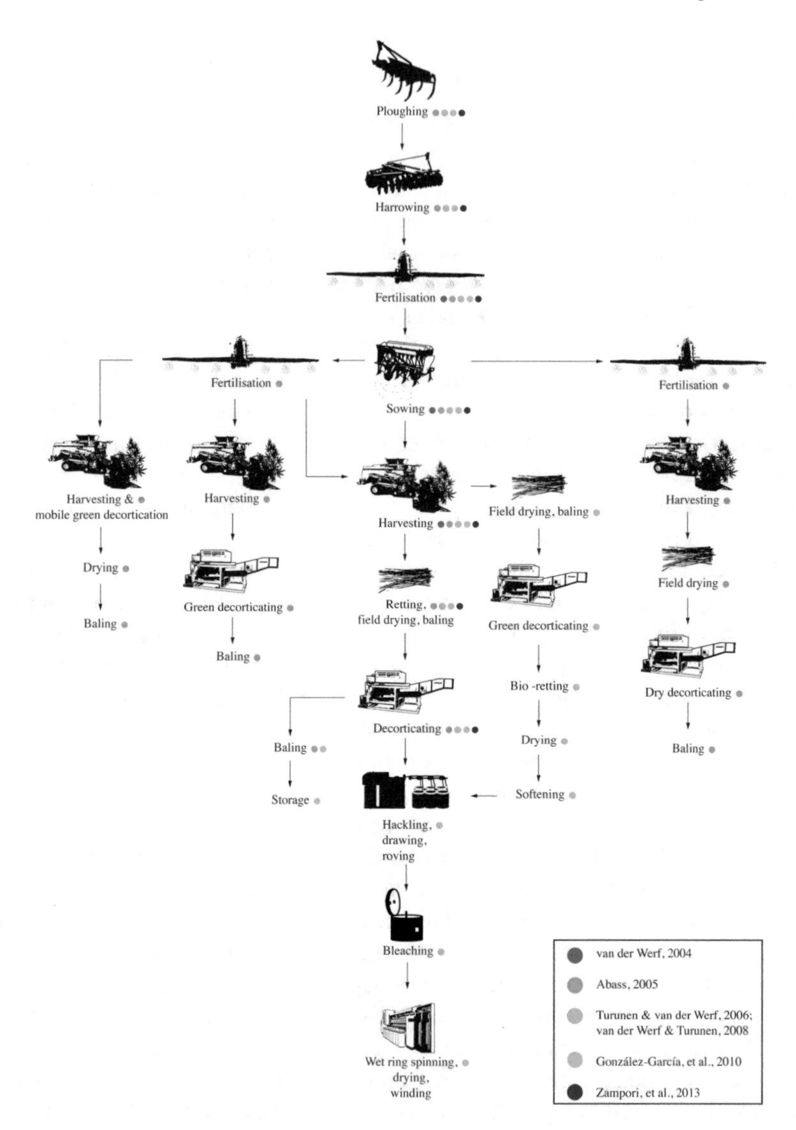

Fig. 2.2 Product life cycle stages of hemp produced in France, UK, Spain, Italy, and Central Europe

assessment of France only includes harvesting and cultivation processes. The LCAs of hemp cultivated in the UK, Italy, and Spain also included decortication of hemp stalks. The LCA in Central Europe was more detailed and also included yarn production processes.

2.3.1 France

The main focus of van der Werf's [42] LCA was to assess the impact of French hemp fibre field production in comparison with other crops. The study assessed the impact according to eutrophication, climate change, acidification, terrestrial ecotoxicity, energy use, and land use. The other crops that were included in this study are sunflower, rapeseed, pea, wheat, maize, potato, and sugar beet.

The values for hemp according to good agricultural practices compared to other crops are consistently low, except for land use. Besides that, the contribution of hemp to eutrophication is high in comparison with overall environmental impacts in Europe. Eutrophication and climate change were predominantly impacted by fertilisation production and field emissions. Acidification was impacted by fertiliser production and field emissions but also by diesel production and use. Energy use was mainly impacted by N-fertiliser production, diesel production and use, and by machinery production [42].

Pig slurry, reduced tillage, and reduced nitrate leaching for hemp production were assessed in comparison with good agricultural practices to analyse the environmental impacts when these production scenarios are implemented. Pig slurry is a natural fertiliser that can replace chemical fertilisers. Tillage is the preparation of land before crops are grown. Reduced tillage decreases the soil disturbance, leaves behind more crop residue on the soil, and decreases erosion. Reduced nitrate leaching can decrease acidification of the soil and reduce the number of chemicals leached into freshwater and marine water [42].

Eutrophication increased with pig slurry, slightly reduced with reduced tillage and was almost cut in half when leaching was reduced. Climate change was lower in all three scenarios compared to good agricultural practices. Acidification stayed the same with reduced leaching, reduced with reduced tillage but increased in the pig slurry scenario. Terrestrial ecotoxicity stayed the same in the reduced tillage and reduced leaching scenarios but increased with pig slurry. The energy use either reduced for pig slurry or reduced tillage, whereas it stayed the same for less leaching. Land use stayed the same in all the production scenarios.

2.3.2 United Kingdom

This LCA was done in the UK by Abass [1] as part of a master's thesis. This thesis includes the following fibre hemp production processes: seed cultivation, hemp cultivations, harvesting, transportation, retting, decortication, drying, and packaging. Several assessments were made in which the decortication methods were distinguished between green, conventional, dry, and mobile green decortication. The four processing methods were analysed according to abiotic depletion, global warming, ozone layer depletion, human toxicity, freshwater aquatic ecotoxicity, marine aquatic ecotoxicity, terrestrial ecotoxicity, photochemical oxidation, acidification, and eutrophication. This LCA did include carbon sequestration of hemp, which resulted in negative results for global warming.

Green decortication is a process in which green stalks are used when the hurds and fibres are separated, as the plants are decorticated right after harvest. Conventional decortication includes field retting. After the stalks are retted, they are decorticated. Dry decortication is the method in which stalks are dried on the field before the stalks are decorticated. The last decortication process of this study, mobile green decortication, is similar to the green decortication process in which green stalks are decorticated. This method distinguishes itself by decorticating the stalks on the field alongside harvesting. The highest fibre yield can be achieved by green and mobile green decortication as the fibre yield is 25% of the total hemp yield [1].

The results of this LCA with regard to the decortication processes showed how the dry decortication method had the best environmental performance. The dry decoration had the lowest impact on global warming. Mobile green decortication was the second-best decortication method with regard to environmental impact. Mobile green decortication scored the lowest on abiotic depletion, freshwater ecotoxicity, marine ecotoxicity, terrestrial ecotoxicity, photochemical oxidation, acidification, and eutrophication. The decoration processes which had the highest contributions to the impact categories are the green decortication method. This method had the largest impact in the six categories ozone layer depletion, freshwater aquatic ecotoxicity, marine aquatic ecotoxicity, terrestrial ecotoxicity, photochemical oxidation, and acidification. Traditional decortication had the largest impact in the category's abiotic depletion, human toxicity, and eutrophication. Overall, drying, the decortication method, and fertiliser use contributed to the largest amount of environmental impact [1].

2.3.3 Spain

Another LCA study, which was based on Spanish cultivation, studied the environmental impacts of hemp and flax as raw materials for non-wood pulps. The long bast fibre extraction of hemp for paper end use and textile end use follows similar paths until the pulp mill. This is a cradle to gate LCA, where the gate is the transport to the pulp mill. Acidification, eutrophication, global warming, photochemical oxidant formation, energy resources, and pesticide use were analysed in this study [17].

For this LCA, hemp scored worse than flax in each category, except for pesticide use, since hemp did not require any pesticides or herbicides. The main contributor to the higher values in this study was the fertiliser use for hemp crops. The fertilisers that were used were nitrogen-based, which lead to N_2O, NH_3, NO_x emissions, and NO_3^- leaching. Other processes with a large impact were irrigation, harvesting, and scutching [17].

2.3.4 Italy

Zampori et al. [47] performed a LCA study on fibre hemp that was cultivated in Italy. Some of the steps that are studied in this LCA study are ploughing, harrowing, pre-sowing fertilisation, sowing, germination, cutting and threshing, windrowing, baling and bales loading, transport to scutching site, scutching, and fibre preparation. This study included the methods greenhouse gas protocol, cumulative energy demand, and eco indicator 99 to carry out the life cycle impact assessment.

Fertilisation had the largest impact on CO_2 equivalent per hectare, followed by ploughing, cutting, and baling. Scutching and transport were not considered with regard to kg of CO_2 eq per ha of cultivated land. If scutching and transport were included, then it would be the second-largest contributor to kg of CO_2 eq per ha in this study, following fertilisation [47].

2.3.5 Central Europe

van der Werf and Turunen [40, 41] conducted a LCA study of hemp and flax yarn production. This study compared several hemp retting scenarios with data gathered from Central Europe and predominantly Hungary and France. The study distinguished hemp water retting, hemp bioretting, baby hemp, and dew retting of flax. Three different production stages were accessed: crop production, fibre processing, and yarn production. Machines, energy use, chemicals, and water use for each stage were included for the analysis. The data were analysed regarding the impact categories eutrophication, climate change, acidification, non-renewable energy use, land occupation, pesticide use, and water use [40, 41].

Eutrophication was the lowest for dew retted flax. The higher outcome for hemp was mainly caused by crop production and emissions from soils and fibre processing which also included retting. Contribution to climate change and non-renewable energy use was the lowest for water retted hemp, followed by dew retted hemp, baby hemp, and bioretted hemp. Bioretted hemp had the highest outcome for climate change and non-renewable energy use due to fibre drying, which was not included in the other scenarios. Acidification was the lowest for water retted hemp, followed by baby hemp and dew retted flax. Bioretted hemp has the largest acidification rate; this was again mainly caused by the process of fibre drying. Furthermore, the land occupation of dew retted flax and water retted hemp was the lowest. In comparison, the land occupation for baby hemp required double the amount of land. Pesticide use for water retted and bioretted hemp was zero, followed by 0.296 kg for dew retted flax and 0.874 kg for baby hemp. The last category regarding water use was the highest for bioretted hemp and water retted hemp, which was caused by the retting and rinsing stage.

Hemp water retted land occupation was slightly higher than dew retted flax. Previous literature mentioned that hemp yields more per hectare than flax [23]. van der Werf and Turunen [41] explained the difference in their study according to the long fibre extraction. Flax had a higher long fibre extraction compared to hemp in this study. The higher fibre extraction compensated for the lower green stem yield of flax. van der Werf and Turunen [41] also mentioned the poor yarn production and fibre processing technology for bast fibres, which resulted in higher energy consumption in these stages compared to cotton [40].

2.4 LCA Hot Spots

Each LCA study was focusing on a different part of hemp's life cycle, whereas one studied the retting processes, another looked at decortication. Some studies only consider hemp crops' production, whereas other studies assessed the whole textile life cycle of hemp. Therefore, the outcome of which hemp fibre or textile production processes contributed the most to the impact categories fluctuated. Crop production had a lower impact on the categories in the studies that include spinning and weaving.

2.4.1 Global Warming Potential

The contribution of hemp to the global warming potential was due to fertiliser use and production [1, 17, 40, 42, 43], decortication [1, 40, 43, 47], wet spinning, yarn drying [40, 43], and weaving [43]. Fibre drying also had a large impact on green decorticated stems [1, 40]. N_2O and CO_2 emissions mainly impact the global warming potential [42]. Fertilisation use and production had a lower contribution to global warming potential in studies that included more production steps for hemp textiles. A change from wet to dry spinning could reduce the global warming potential, as wet spinning requires more energy and includes an additional drying process. Furthermore, the use of more efficient and newer weaving machinery could also reduce global warming potential [43].

2.4.2 Terrestrial Acidification

Terrestrial acidification of hemp textile production is due to fertiliser use, fertiliser production [1, 17, 40, 42, 43], decortication [1, 17, 42], spinning [40, 43], and weaving [43]. Fibre drying also had a substantial impact on acidification for dry, green, and mobile green decorticating [1, 40]. Around one-third of hemp production's terrestrial acidification potential is due to fertiliser use and production [43]. NH_3, SO_2, and NO_2 emissions contribute to terrestrial acidification [42]. Mobile green decortication [1] and dry spun hemp [43] had the lowest acidification rate.

2.4.3 *Eutrophication*

Most of the studies assessed contribution to eutrophication as a whole rather than assessing marine and freshwater separately. These studies considered fertilisation use and the accompanying nitrogen and phosphorus soil emissions as a large contributor to eutrophication [1, 17, 40, 42] and a small fraction due to retting and yarn production [40]. Van Eynde [43] separated the eutrophication of freshwater and marine. Eutrophication of freshwater was due to the degumming or bleaching process, and only a small fraction was due to fertiliser use and production. Marine eutrophication was mainly due to fertiliser use and production.

2.4.4 *Abiotic Depletion*

Abiotic depletion in hemp fibre production for conventional, dry or green decortication was due to decortication, fertilisation, and hemp seed production. Abiotic depletion for hemp that was mobile green decorticated was due to fertilisation use and seed production [1].

2.4.5 *Ozone Depletion*

Conventional and dry decorticated hemp fibre contribute to ozone layer depletion through decortication, land preparation, fertiliser use, retting, and sowing. Ozone layer depletion for mobile green and green decorticated hemp is due to fibre drying, decortication, and land preparation [1].

2.4.6 *Photochemical Oxidation*

Hemp fibre production contributes to photochemical oxidation through fertiliser use, decortication [1, 17], and land preparation. The dying process of green and mobile green decortication is an extra factor that increases photochemical oxidation [1].

2.4.7 Ecotoxicity

Field emissions, such as fertilisation, sowing, and land preparation, contribute to a larger share of terrestrial ecotoxicity [1, 42]. Furthermore, decortication and fibre drying also had a share of terrestrial ecotoxicity [1]. Terrestrial ecotoxicity was due to Ni and Cd emissions [42]. Freshwater ecotoxicity was due to crop fertilisation, land preparation sowing, fibre drying, and decortication [1, 43]. For bleached hemp, sizing and weaving also contributed to freshwater ecotoxicity. Marine ecotoxicity is due to sizing and weaving, fibre preparation, spinning, degumming, and bleaching. Hemp crop production only had a small share of this impact category [43].

2.4.8 Human Toxicity

Human toxicity is due to land preparation, decortication, sowing, fertilisation, retting, harvesting, fibre preparation, spinning, sizing, and weaving [1, 43]. The decortication processes, mobile green and green, also influence human toxicity through fibre drying [1]. Bleached hemp has a higher contribution than decorticated hemp due to more extensive sizing and weaving emissions [43].

2.4.9 Particulate Matter Formation

Particulate matter formation is due to fibre preparation, spinning, weaving, degumming, and bleaching. The crop production of hemp has a relatively small contribution to this category, with only a small number of emissions from N-fertiliser use [43].

2.4.10 Energy Use

The contribution of hemp to energy use was due to crude oil and natural gas use. N-fertiliser production, diesel production and use, machinery production, yarn production, and fibre and yarn drying contribute to energy use [17, 40, 42]. Yarn production contributed the most to energy use, followed by fibre drying for bioretted hemp [40].

2.4.11 Land Use

The land occupation of hemp was mainly due to the crop production phase. Efficient fibre use can decrease the amount of land needed for hemp fibre production [43]. The contribution of hemp to land occupation is mainly essential in high densely populated countries [42].

2.4.12 Climate Hot Spots

The outcome of the studies with regard to what should be improved about the current fibre hemp production processes mainly mentioned the need to improve the fertilisation, decorticating, spinning, and drying phases.

Changing the type of fertiliser could reduce the emissions from fertiliser use and production [17]. More efficient use of fertilisers would also reduce the environmental impact of hemp and be an important economic improvement for farmers. To improve the environmental impact of hemp with regard to eutrophication and global warming potential, reduced nitrate leaching is suggested by van der Werf [42]. Nitrogen leaching can be reduced by optimised fertilisation and reduced time in-between two crop growth periods. The optimisation of fertilisation use can be changed by improved genotypes [42].

As the decortication process is inefficient, an improvement of this would influence the environmental impact of hemp [43]. Furthermore, a reduction in the use of fossil fuels and increased use of renewable energy could reduce the impacts of decortication [1]. To improve sustainability, breeding of high quality and biomass crops is needed, combined with improved fibre separation techniques and the use of more renewable energy [43].

The spinning technology of bast fibres is currently underdeveloped and more complicated compared to techniques for other fibres. Another cause of the large contribution of spinning to the impact categories is that bast fibres spinning is less efficient, which results in higher energy use [40]. Energy use can be improved by using renewable energy sources and improved spinning and other yarn technologies for bast fibres [43].

Drying was, in some cases, part of the decortication step, such as for green and mobile green decortication [1]. It is also part of the wet spinning process, where the yarns are dried after spinning [40, 43]. A change in the decortication method can remove the need for fibre drying and reduce the accompanying emissions and impact of this production process [1, 40]. Besides that, hemp can be dry spun, which does not require yarn drying [43].

2.4.13 China Versus Europe

A difference between hemp produced in China or Europe could be the use of fertilisers. Overall, the use of fertilisers in China is less efficient than in Europe [43]. The amount of fertiliser use does not seem to vary significantly in Europe [40]. European grown hemp currently has a lower environmental impact than hemp that is grown in China due to this difference in fertiliser use. Furthermore, the degumming process in Europe seemed to have a lower environmental impact compared to the decortication process in China. This is mainly due to the difference in energy use, where coal is still a large energy source in China, it is not in Europe [43].

2.5 Hemp Versus Other Textile Fibres

Hemp is often seen as a sustainable alternative for the currently most used fibres. The following studies assessed hemp compared to cotton, flax, and polyester.

The first study of Cherrett et al. [7] compares twelve scenarios of hemp, organic hemp, cotton, organic cotton, and polyester according to energy requirements, carbon dioxide emissions, ecological footprint, and water analysis. The fibres were further divided into production into India, Europe, and USA, and between traditional processing, semi-experimental, and experimental processing methods. The experimental process for hemp was green decorticated stalks combined with chemical degumming. Dew retted hemp in combination with chemical degumming was described as a semi-experimental process.

The energy requirements were divided between crop cultivation and fibre production. Polyester has the highest energy use in total. The crop cultivation was the lowest for organic hemp and the highest for cotton produced in the USA. The energy requirements for cotton and hemp fluctuate for crop cultivation depending on the country; the fibre was cultivated in and the processing methods. Fibre processing for hemp had higher results than cotton. Especially the experimental processing methods of hemp required higher energy use. The total results of crop cultivation and fibre production combined resulted in the lowest result for organic cotton and organic hemp with traditional processing methods and the highest for hemp processed with experimental or semi-experimental processing methods.

Carbon dioxide emissions were also separated for crop cultivation and fibre cultivation. The lowest results for carbon dioxide emissions were for organic cotton and organic hemp. The highest carbon dioxide emissions were for polyester. In general, the fibre production processes for hemp had higher carbon dioxide emission than cotton. In total, cotton produced in the USA, and experimental hemp had, besides polyester, the highest carbon dioxide emissions.

The summary of results for ecological footprint resulted in the lowest footprint for organic and traditional produced hemp, the highest for cotton, organic cotton, and polyester. Experimental processing methods for hemp had a larger footprint than traditional hemp processing methods.

Turunen and van der Werf's [40] study shows the different environmental impacts of hemp and flax. As mentioned earlier, this study included different hemp scenarios: baby hemp, water retted hemp, and bioretted hemp. Water retted hemp and flax are very similar except for eutrophication, water use, and pesticide use. Eutrophication was slightly higher for hemp due to retting effluents. Water use was much higher for water retted hemp as the water is also used for retting. However, this water can be reused, which would lower the eventual water. Pesticide use, on the other hand, is higher for flax as hemp does not need pesticides.

Turunen and van der Werf [40] also mention that hemp requires more land for 100 kg of yarn than cotton and flax. This is caused by the lower long fibre output of hemp compared to cotton and flax. However, as every part of the hemp crop can be used, the outcome of land use in comparison with the final long fibre output should be put into perspective.

2.6 End-of-Life

There are multiple ends of life possible for hemp textile or apparel. Some common things for consumers or brands could be either repair, resell, redesign, or upcycle the item. Some companies are taking back old clothes from consumers to resell or to recycle. Another possibility is upcycling of garments and turning them into entirely new products. Apparel should only be recycled or biodegraded when reuse is not possible anymore; this could be the case if apparel is washed too often or damaged in a way that the textile cannot be repaired [11].

There are currently two types of textile recycling possibilities, mechanical recycling, and chemical recycling. The first method is a downcycling method where the textiles cut to the preferred size for cleaning wipes, and the machine shreds fibres for insulation or fibre-to-fibre recycling. Shredded fabrics for fibre-to-fibre recycling are separated by colour. After this, the textiles are shredded, carded, and spun. As the fibres become short due to the shredding procedure, virgin material is added to ensure the final product's strength. Around 40% of the fibres are lost in the mechanical recycling process [29, 34]. The second method, chemical recycling, is a fibre-to-fibre recycling method. The end product of natural fibres recycled through a chemical process is similar to lyocell or viscose fibres [25]. Chemical recycled fibres are strong enough and do not need to be mixed with virgin material [37]. Depending on the chemicals used, the recycling process does not impact fibres' colourfastness, and thus fibres retain their original colour [29].

As hemp is a natural fibre, it is also fully biodegradable [31]. Consumers can compost their old hemp clothes if they are not suitable for recycling or reuse in the compost bin, where the textile eventually will biodegrade. Composting of textiles can have some hazards as additional dyes are often used in textile processes [11]. The cradle-to-cradle certified product standard was developed to ensure that textiles can be safely composted as they are required to be made of safe materials and chemicals before the apparel items can be certified [8]. A study of 2019 also studied the possibility of using textile waste, combined with other waste, as organic fertilisers, which might be a possibility in the future [4].

2.7 Social Sustainability

Social sustainability is the community's ability to create configurations to sustain a healthy society and support future generations without comprising existing constituents [18]. To assess the social sustainability of hemp textile production, production hazards and sustainable work environments are reviewed.

2.7.1 Production Hazards

Workers from textile producing facilities come into contact with a lot of hazards. The most common hazards are mechanical hazards, chemical hazards, biological agent hazards, ergonomic hazards, and psychological hazards [18].

To start with the mechanical hazards, as workers are working with fibre processing machines, they get getting into contact with hemp dust. The organic dust of the agricultural and textile workers is exposed to can cause serious health issues, depending on the dust mixture. The dust can contain plants particles, glucans, viruses, bacteria and endotoxin, fungi and mycotoxins, pollen, insects, and compost. Inhalation of organic dust can cause respiratory infections, irritation, inflammation, allergies, toxic response, or a combination of these [10]. Studies have identified the presence of the bronchoconstrictive disease, byssinosis, among hemp workers [6, 12, 50, 51]. Byssinosis often goes together with chest tightness, fever, headache, or muscle aches [6]. High exposure to inhalable organic dust is a serious health risk in hemp production processes [9]. If organic dust levels in hemp processing units are too high, general and local ventilation are recommended. Furthermore, to protect the health of the worker's respirators should be provided according to the Occupational Safety and Health Administration [33].

Workers are also exposed to certain chemical hazards such as the use of pesticides, herbicides, or fungicides [26]. Even though hemp does not require the use of pesticides and herbicides [48], certain farmers still choose to use these chemicals. The use of fertilisers that promote plant growth is more common [1, 17, 40–43], but only require a low input [48]. Workers that get into contact with these chemicals can suffer serious health issues and chronic diseases [5], including acute and chronic neurotoxicity [36]. There has also been a link between certain pesticides and cancer [3]. These hazards do not apply to hemp that is grown organically without the use of herbicides, pesticides, and inorganic fertilisers [9]. On some farm's hemp is grown indoor. Carbon dioxide is sometimes added to promote the growth of the crop in indoor farms. Exposure to an excess of carbon dioxide can create oxygen deficiencies in the atmosphere. The use of carbon dioxide also exposes the workers to hazards of handling compressed gasses [26]. Textile processes also contribute to chemical hazards for workers. Especially workers that work in dyeing, printing, and finishing are exposed to hazardous chemicals. Possible hazards for workers due to chemical use are skin and eye irritation, lung oedema, burns, and DNA mutation. The hazard to textile workers depends on the way that workers are exposed to the chemicals [38]. Besides the workers that are affected by the use of chemicals, local populations can also be affected through the spread of these chemicals in contaminated water [36]. Due to these risks, more initiatives focus on protecting workers health and the environment. One example is the zero discharge of hazardous chemicals (ZDHC), which focuses on reducing hazards for chemicals that are used in textile facilities [49].

Some biological hazards that relate to hemp crop production are allergies and breathing disorders. Several studies found the possibilities of an allergic skin reaction or allergic rhinitis to hemp pollen that were dispersed in the atmosphere. The allergic reaction to hemp pollen was often paired with asthma-like symptoms [16, 24, 39, 45]. The release of pollen is mainly during summer and autumn [2]. Exposure to decorticating hemp in combination with organic dust has also led to allergic reactions in some cases [15]. Organic dust can worsen allergic reactions [30].

Ergonomic hazards are hazards related to the physical risk that can cause musculoskeletal disorders (MSDs). The heavy labour of the textile industry is a reason for MSDs among textile workers [18]. Traditional hemp processing is labour intensive and, therefore, is only carried out in countries with low labour cost, such as Eastern Europe, or Asia [40]. In these countries, workers are more prone to ergonomic hazards [18]. Physical factors that lead to ergonomic hazards and can cause injury are constant posture, force [28], noise, repetition, working environment, and vibration. Ergonomic risk can be reduced through proper work techniques, avoid excessive force, job rotations, training, and education [35].

2.7.2 Sustainable Work Conditions

Sustainability of the work conditions includes human rights, basic necessities of workers, fair wages, and work time [18].

Some of the largest problems in the textile industry, related to human rights, are unfavourable work conditions, discrimination, prohibition of association, and child and forced labour [20]. Discrimination among workers is another problem of the textile industry. A large number of textile workers are unqualified female workers [21]. Young female workers are sometimes employed with the intention to exploit them. In some countries, females have a lower ranking than man, which increases the exploitation of women [19]. The freedom of association is also often denied. It is more difficult to form associations for textile workers as a lot of work is informal and done by homeworkers in combination with irregular work times and unstable work relations [20, 46]. Child labour is another issue that is also widespread among the textile and clothing sector. As the wages for female workers are often low, they involve their children in their work to sustain themselves [13]. Child labour is banned in a lot of countries, and the labour can violate the right of children to education. As these children are not educated, their chances to improve their future perspective and living conditions are lowered [20].

The textile and fashion industries are often related to low wages. Due to this, a lot of workers are living in poverty with incomes that are insufficient to support their family and basic necessities [20]. According to International Labour Office (ILO) [21], the low wages are among others caused by the high competition in the industry and the composition of the workforce. Poor wages are often combined with high workload and overtime. As workers do not earn enough, they work extra time to get additional pay. Overtime is furthermore a result of poor administration, unskilled employees that need more work for certain tasks, excessive consumerism, and consumer trends that increase the workload for factories [18].

More research about the social sustainability of hemp is needed to understand the impact that hemp crop and textile production have on local communities and if the fibre crop contributes to a healthy community.

2.8 Conclusion

Hemp crop cultivation has a low environmental impact compared to other crops. The fibre processing, however, diminishes its sustainable nature and increasing the ecological impact of hemp. The main causes for this are fertilisation use, decortication processes, spinning, and drying. Either a change of fertilisation use or more efficient use of fertilisers could already improve hemp's sustainable performance. The decortication phase is a fibre processing step that is only applicable for bast fibres. Therefore, other natural fibres such as cotton do not have this extra process. The use of renewable energy during the decortication stage could decrease emission. The same applies to the spinning operation; if the industry used more renewable energy in the future, the emissions from this process would decrease. Next to that, spinning processes for hemp fibres are less efficient than spinning for cotton. The last process that increases hemp's environmental impact was the drying process. As hemp fibres are often wet spun, they also need to be dried. The drying process is an energy-intensive process. To remove the need for drying after spinning, hemp can also be dry spun. Another drying process is part of some decortication methods of hemp. A change of the decortication method can reduce the need for an additional drying process. Currently, hemp biomass harvest is higher than cotton and flax, but the eventual kilogram of yarn that can be spun of one hectare is much lower for hemp. This lower yarn output is due to a lower number of long fibres from the hemp harvest. Improved fibre processing methods and a change in genotype could increase the eventual long fibre output of hemp per hectare.

The social sustainability of hemp showed quite some production hazards and possible work environment problems common in the textile industry. To better assess the social sustainability of hemp, monitoring should improve.

References

1. Abass E (2005) Life cycle assessment of novel hemp fibre–a review of the green decortication process. Imperial College London, Department of Environmental Science and Technology, London
2. Abbas S et al (2012) World allergy organization study on aerobiology for creating first pollen and mold calendar with clinical significance in Islamabad, Pakistan; a project of world allergy organization and Pakistan allergy, asthma & clinical immunology Centre of Islamabad. World Allergy Organ J 5(9):103–110. https://doi.org/10.1097/WOX.0b013e31826421c8
3. Alavanja MCR, Bonner MR (2012) Occupational pesticide exposures and cancer risk: a review. J Toxicol Environ Health 15(4):238–263. https://doi.org/10.1080/10937404.2012.632358
4. Biyada S, Merzouki M, Imtara H, Benlemlih M (2019) Assessment of the maturity of textile waste compost and their capacity of fertilization. Eur J Sci Res 154(3):399–412
5. Blair A, Ritz B, Wesseling C, Beane Freeman L (2014) Pesticides and human health. Occup Environ Med 72(2):81–82. https://doi.org/10.1136/oemed-2014-102454
6. Bouhuys A et al (1967) Byssinosis in hemp workers. Arch Environ Health Int J 14(4):533–543. https://doi.org/10.1080/00039896.1967.10664790
7. Cherrett N et al (2005) Ecological footprint and water analysis of cotton, hemp and polyester. Stockholm Environment Institute, Stockholm. Available at: https://mediamanager.sei.org/documents/Publications/SEI-Report-EcologicalFootprintAndWaterAnalysisOfCottonHempAndPolyester-2005.pdf. Accessed 3 Feb 2021
8. Cradle to Cradle Product Innovation Institute (2016) Cradle to cradle certified—product standard version 3.1. Cradle to Cradle Product Innovation Institute, Oakland
9. Davidson M et al (2018) Occupational health and safety in cannabis production: an Australian perspective. Int J Occup Environ Health 1–11. https://doi.org/10.1080/10773525.2018.1517234
10. Donham KJ, Thelin A (2016) Agricultural respiratory diseases. In: Donham KJ, Thelin A (eds) Agricultural medicine–rural occupational and environmental health, safety and prevention. Wiley, Hoboken, pp 95–149
11. Ellen MacArthur Foundation (2017) A new textiles economy: redesigning fashion's future. Ellen MacArthur Foundation, Cowes. Available at: https://www.ellenmacarthurfoundation.org/assets/downloads/publications/A-New-Textiles-Economy_Full-Report.pdf. Accessed 10 Feb 2021
12. Er M et al (2016) Byssinosis and COPD rates among factory workers manufacturing hemp and jute. Int J Occup Med Environ Health 29(1):55–68. https://doi.org/10.13075/ijomeh.1896.00512
13. Ergon (2008) Literature review and research evaluation relating to social impacts of global cotton production for ICAC expert panel on social, environmental and economic performance of cotton (SEEP). Ergon, London. Available at: https://ergonassociates.net/wp-content/uploads/2011/06/literature_review_july_2008.pdf. Accessed 11 Feb 2021
14. FAOSTAT (2020) Crops hemp tow waste. Available at: http://www.fao.org/faostat/en/#data/QC. Accessed 05 Feb 2021
15. Fishwick D et al (2001) Respiratory symptoms, lung function and cell surface markers in a group of hemp fiber processors. Am J Ind Med 39(4):419–425. https://doi.org/10.1002/ajim.1033

16. Freeman GL (1983) Allergic skin test reactivity to marijuana in the southwest. West J Med 138(6):829–831. Available at: https://www.ncbi.nlm.nih.gov/pmc/articles/PMC1010828/pdf/westjmed00202-0047.pdf. Accessed 11 Feb 2021
17. González-García S, Hospido A, Feijoo G, Moreira MT (2010) Life cycle assessment of raw materials for non-wood pulp mills: hemp and flax. Resour Conserv Recycl 54(11):923–930. https://doi.org/10.1016/j.resconrec.2010.01.011
18. Grace Annapoorani S (2016) Social sustainability in textile industry. In: Muthu S (ed) Sustainability in the textile industry. Springer, Singapore, pp 57–78
19. Hale A, Wills J (2007) Women working worldwide: transnational networks, corporate social responsibility and action research. Glob Netw 7(4):453–476. https://doi.org/10.1111/j.1471-0374.2007.00179.x
20. Hamm B (2012) Challenges to secure human rights through voluntary standards in the textile and clothing industry. In: Cragg W (ed) Business and human rights. Edward Elgar Publishing, Cheltenham, pp 220–242
21. ILO (2005) Promoting fair globalization in textiles and clothing in a post-MFA environment. ILO, Geneva. Available at: https://www.ilo.org/wcmsp5/groups/public/—ed_dialogue/—sector/documents/meetingdocument/wcms_161673.pdf. Accessed 11 Feb 2021
22. ISO (2006) ISO 14040:2006 environmental management—life cycle assessment—principles and framework. International Organization for Standardization—ISO, Geneva
23. Jordan HV, Lang AL, Enfield GH (1946) Effects of fertilizers on yield and breaking strengths of American hemp, *Cannabis sativa*. J Am Soc Agron 38(6):551–563. https://doi.org/10.2134/agronj1946.00021962003800060010x
24. Kumar R, Gupta N (2013) A case of bronchial asthma and allergic rhinitis exacerbated during Cannabis pollination and subsequently controlled by subcutaneous immunotherapy. Indian J Allergy Asthma Immunol 27(2):143–146. https://doi.org/10.4103/0972-6691.124399
25. Ma Y, Zeng B, Wang X, Byrne N (2019) Circular textiles: closed loop fiber to fiber wet spun process for recycling cotton from denim. ACS Sustain Chem Eng 7:11937–11943. https://doi.org/10.1021/acssuschemeng.8b06166
26. Martyny JW, Serrano KA, Schaeffer JW, van Dyke MV (2013) Potential exposures associated with indoor marijuana growing operations. J Occup Environ Hyg 10(11):622–639. https://doi.org/10.1080/15459624.2013.831986
27. Mcgrath C (2020) 2019 hemp annual report—Peoples Republic of China. USDA, Washington D.C. Available at: https://apps.fas.usda.gov/newgainapi/api/Report/DownloadReportByFileName?fileName=2019%20Hemp%20Annual%20Report_Beijing_China%20-%20Peoples%20Republic%20of_02-21-2020. Accessed 7 Nov 2020
28. Nagaraj TS, Jeyapaul R, Vimal KEK, Mathiyazhagan K (2019) Integration of human factors and ergonomics into lean implementation: ergonomic-value stream map approach in the textile industry. Prod Plann Control 30(15):1265–1282. https://doi.org/10.1080/09537287.2019.1612109
29. Nasri-Nasrabadi B, Wang X, Byrne N (2020) Perpetual colour: accessing the colourfastness of regenerated cellulose fibres from coloured cotton waste. J Text Inst 111(12):1745–1754. https://doi.org/10.1080/00405000.2020.1728182
30. Nayak AP, Green BJ, Sussman G, Beezhold DH (2017) Allergenicity to *Cannabis sativa* L. and methods to assess personal exposure. In: Chandra S, Lata H, Elsohly MA (eds) *Cannabis sativa L.—botany and biotechnology*. Springer, Cham, pp 263–284
31. Netravali AN (2005) Biodegradable natural fiber composites. In: Blackburn RS (ed) Biodegradable and sustainable fibres. Woodhead Publishing, Abington, pp 271–309
32. Okubo M, Kuwahara T (2020) Emission regulations. In: Okubo M, Kuwahara T (eds) New technologies for emission control in marine diesel engines. Butterworth-Heinemann, Oxford, pp 25–51
33. OSHA (2011) 1910.134—respiratory protection. Available at: https://www.osha.gov/laws-regs/regulations/standardnumber/1910/1910.134. Accessed 10 Feb 2021

34. Östlund Å et al (2015) Textilåtervinning. Tekniska möjligheter och utmaningar. Naturvårdsverket, Stockholm. Available at: https://docplayer.se/4277866-Textilatervinning-tekniska-mojligheter-och-utmaningar.html. Accessed 11 Feb 2021
35. Polat O, Kalayci CB (2016) Ergonomic risk assessment of workers in garment industry. Textile Science and Economy VIII, Zrenjanin
36. Sankhla MS et al (2018) Water contamination through pesticide & their toxic effect on human health. Int J Res Appl Sci Eng Technol 6(1):967–970. https://doi.org/10.22214/ijraset.2018.1146
37. Schuch AB (2016) The chemical recycle of cotton. Revista Produção e Desenvolvimento 2(2):64–76. https://doi.org/10.32358/rpd.2016.v2.155
38. Srother JM, Niyogi AK (1998) Dyeing, printing and finishing. In: Stellman JM, Ivester AL, Neefus JD (eds) Encyclopaedia of occupational health and safety, vol 3, 4th edn. International Labour Organization, Geneva, pp 89.17–89.19
39. Stokes JR, Hartel R, Ford LB, Casale TB (2000) Cannabis (hemp) positive skin tests and respiratory symptoms. Ann Allergy Asthma Immunol 85(3):238–240. https://doi.org/10.1016/S1081-1206(10)62473-8
40. Turunen L, van der Werf HMG (2006) Life cycle analysis of hemp textile yarn. Comparison of three fibre processing scenarios and a flax scenario. INFRA, UMR SAS, Rennes
41. van der Werf HMG, Turunen L (2008) The environmental impacts of the production of hemp and flax textile yarn. Ind Crops Prod 27(1):1–10. https://doi.org/10.1016/j.indcrop.2007.05.003
42. van der Werf HMG (2004) Life cycle analysis of field production of fibre hemp, the effect of production practices on environmental impacts. Euphytica 140(1):13–23. https://doi.org/10.1007/s10681-004-4750-2
43. Van Eynde H (2015) Comparative life cycle assessment of hemp and cotton fibres used in Chinese textile manufacturing. KU Leuven, Leuven
44. Vollenweider RA (1992) Coastal marine eutrophication: principles and control. In: Vollenweider RA, Marchetti R, Viviani R (eds) Marine coastal eutrophication. Elsevier Science, Amsterdam, pp 1–20. https://doi.org/10.1016/b978-0-444-89990-3.50011-0
45. Williams C, Thompstone J, Wilkinson M (2007) Work-related contact urticaria to *Cannabis sativa*. Dermatitis 58(1):62–63. https://doi.org/10.1111/j.1600-0536.2007.01169.x
46. Wills J, Hale A (2005) Threads of labour in the global garment industry. In: Hale A, Wills J (eds) Threads of labour. Blackwell Publishing Ltd., Malden, pp 1–15
47. Zampori L, Dotelli G, Vernelli V (2013) Life cycle assessment of hemp cultivation and use of hemp-based thermal insulator materials in buildings. Environ Sci Technol 47(13):7413–7420. https://doi.org/10.1021/es401326a
48. Zatta A, Monti A, Venturi G (2012) Eighty years of studies on industrial hemp in the Po Valley (1930–2010). J Nat Fibers 9(3):180–196. https://doi.org/10.1080/15440478.2012.706439
49. ZDHC (2019) ZDHC impact report. Available at: https://www.roadmaptozero.com/impact-report#How-We-Started. Accessed 11 Feb 2020
50. Zuskin E et al (1990) Respiratory symptoms and lung function in hemp workers. Occup Environ Med 47(9):627–632. https://doi.org/10.1136/oem.47.9.627
51. Zuskin E, Mustajbegovic J, Schachter EN (1994) Follow-up study of respiratory function in hemp workers. Am J Ind Med 26(1):103–115. https://doi.org/10.1002/ajim.4700260109

Chapter 3
Optimal Conditions for Hemp Fibre Production

3.1 Differences Between Hemp Cultivars

Fibre hemp can be divided into various genotypes and cultivars [16]. These genotypes are adapted to different weather conditions and are heavily affected by climate conditions. Some cultivars thrive under long days, whereas other cultivars flourish under shorter days. The same goes for the preferred temperature and precipitation [40]. In general, fibre hemp is adapted to multiple climate conditions, but the plants grow best in the temperate zone [19]. Hemp grows best in cooler, more moist climates than in zones with high temperatures [7]. Temperate climate zones are defined as environments with moderate precipitation either in the summer, the winter, or throughout the year. Wet summer or winter periods are combined with drier seasons. The summers vary from cold and warm to hot, and the coldest months are above 0 °C [39].

Hemp can be categorised according to several features including population type, plant use, flowering time and gender, and geographic origin [42].

- The population type of hemp is distinguished by wild and naturalised populations, landraces, and cultivars.
- Plant uses of hemp can vary between fibre cultivars used for pulp or long fibres, seed cultivars, drug strains, and ornamentals [16].
- Flowering types are divided in early ripening, later ripening, and intermediate ripening stains. The gender type of hemp is either monoecious or dioecious [42].
- Geographic distribution of the crop resulted in these in these different attributions. A study of the Dutch botanist De Meijer [16] divided the geographic location of hemp into four eco-geographical groups: Northern ecotypes, Central ecotypes, Southern ecotypes, and Far-Eastern ecotypes. As shown in Fig. 3.1, the dispersal of hemp started from the origin of hemp in Central Asia [27] to Northern regions (Northern Russia, Finland), Central regions (Central Russian, Ukraine), Southern regions (Mediterranean region, Balkan, Turkey, Caucasus), and Far-Eastern regions (China, Japan, and South Korea) [16].

F. Dhondt and S. S. Muthu, *Hemp and Sustainability*, Sustainable Textiles: Production, Processing, Manufacturing & Chemistry,
https://doi.org/10.1007/978-981-16-3334-8_3

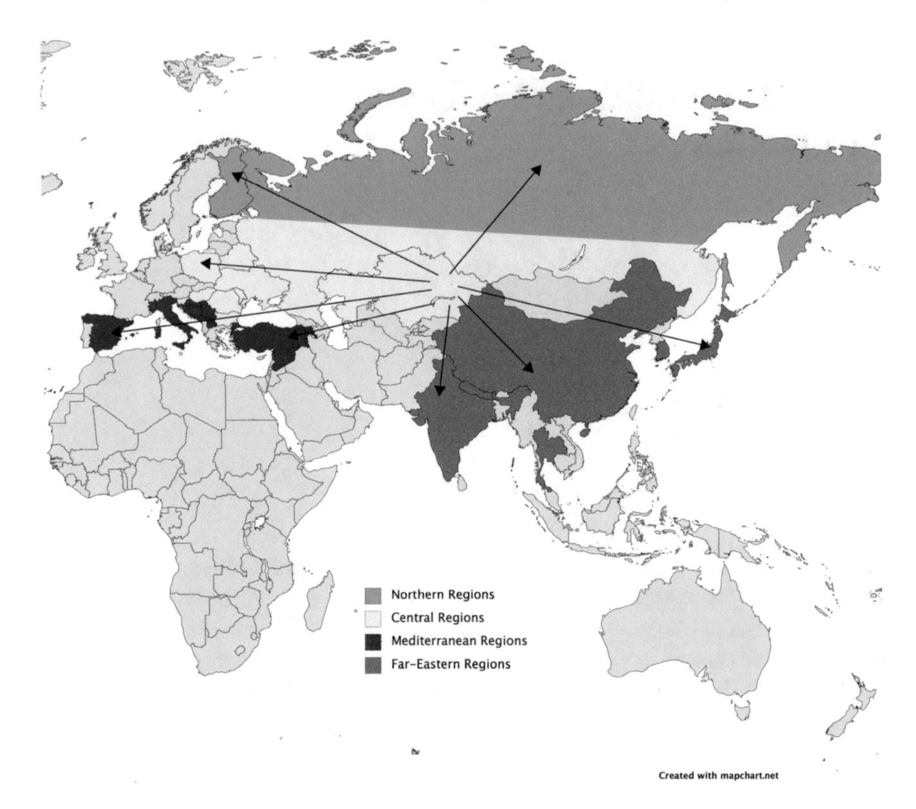

Fig. 3.1 Hemp dispersal by humans from the origin of *Cannabis sativa* L. in Central Asia

Besides the hemp genotypes that were dispersed to other regions and gradually adapted to those environments, humans also bred cultivars for specific environments and end-uses. Crop breeding is the practice in which superior phenotypes are created, selected, and fixated to develop improved hemp cultivars [5]. As soon as the large impact of the environment on yield and quality of fibre hemp was defined [40, 42], breeding programmes started to improve its methods selection criteria accordingly [41].

The study of Petit et al. [40] analysed 123 hemp accessions according to the interactions between genotype and environment. One of the results of this study was that hemp traits are very adaptable to different environmental conditions. The environment affects the length, stem width, and flowering of plants. Besides that, the number of pectins in hemp was also different per location. Furthermore, the environment's impact differs per genotype, and some cultivars are more sensitive to certain environments.

There are currently around 75 hemp cultivars, which are allowed to grow in Europe [20] and 56 that are allowed to grow in Canada [23]. These numbers include both hemp cultivars that are grown for fibre, seed, CBD, or are dual purpose.

Table 3.1 displays an overview of studied fibre genotypes. Table 3.1 shows the correlation of the origin, ecotype (1), sexual type (2), and maturity (3).

1. The ecotypes are divided according to the four regions that De Meijer [16] defined. Canadian cultivars that were not included in De Meijer's study are placed under the Central-Northern regions. The climatic conditions in Canada regions are similar to those regions according to the Köppen-Geiger climate classification [39]. The cultivars are bred to be adapted to the regions mentioned in Table 3.1. However, some cultivars grow even better in other regions than in the proposed ecotypes, as shown in the studies below.

Table 3.1 Overview of the origin, sexual type, climate, and maturity of various hemp cultivars

Reference	Cultivar	Origin	Ecotype	Sexual type	Maturity[a]
Johnson et al. [29]	Alyssa	CA	Central-North	Monoecious	L
Cosentino et al. [14]	Beniko	PL	North	Monoecious	E
Cosentino et al. [14]	Bialobrzeskie	CZ/PL	Central-North	Monoecious	E–I
Cosentino et al. [14]	Carmagnola	IT	Central-South	Dioecious	L
Cosentino et al. [14]	Chamaeleon	NL	Central-North	Dioecious	E
Johnson et al. [29]	CFX-1	CA	Central-North	Dioecious	L
Johnson et al. [29]	CFX-2	CA	Central-North	Dioecious	L
Johnson et al. [29]	CRS-1	CA	Central-North	Dioecious	L
Zatta et al. [58]	CS	IT	Central-South	Dioecious	L
Johnson et al. [29]	Delores	CA	Central-North	Monoecious	L
Cosentino et al. [14]	Dioica 88	FR	Central-South	Dioecious	L
Zatta et al. [58]	Eletta Campana	IT	Central-South	Dioecious	L
Cosentino et al. [14]	Epsilon 68	FR	Central-North	Monoecious	I
Cosentino et al. [14]	Fedora 17	CH/FR	Central-North	Monoecious	E
Zatta et al. [58]	Fedrina 74	FR	Central-North	Monoecious	E
Cosentino et al. [14]	Felina 34	FR	Central-North	Monoecious	E
Cosentino et al. [14]	Ferimon	FR	Central-North	Monoecious	E
Cosentino et al. [14]	Fibranova	IT	Central-South	Dioecious	L
Zatta et al. [58]	Fibrimon 56	FR	Central-North	Monoecious	E
Cosentino et al. [14]	Futura 75	FR	Central-North	Monoecious	I
Zatta et al. [58]	Kompolti hybrid TC	HU	Central-North	Dioecious	I
Cosentino et al. [14]	Lovrin 110	RO	Central	Dioecious	I–L
Cosentino et al. [14]	Tiborszallasi	HU/IT	Central	Dioecious	I–L
Li et al. [33]	Yunma 2	CN	Far Eastern	Dioecious	L
Li et al. [33]	Yunma 3	CN	Far Eastern	Dioecious	L
Li et al. [33]	Yunma 4	CN	Far Eastern	Dioecious	L

[a]E = early (40–60 days); I = intermediate (60–90 days); L = late (90–120 days)

Tang et al. [50] studied 14 different hemp cultivars in four different locations. The highest latitude was at 57° with a maximum daylength of 17.5 h in Latvia, and the lowest latitude was 45° with maximum daylength of 15.4 h in Italy. The mean temperature in Latvia was 16.3 °C whereas in Italy it was 21.6 °C. The other latitudes of 50° in the Czech Republic and 48° in France had a daylength of 16.1 and 15.9 h. The mean temperature in the Czech Republic and France was 16 and 17.7 °C. Precipitation for the four countries was 216.4 mm in Latvia, 260.4 mm in the Czech Republic, 330.6 in France, and 131.4 in Italy. The best performing locations for hemp bast fibre yield were Latvia and the Czech Republic. The cultivars Beniko, Bialobrzeskie, CS, Epsilon 68, Ferimon, and Tiborszallasi performed best in Latvia, whereas the cultivars Fedora 17, Felina 32, and Futura 75 performed best in the Czech Republic. The growth of hemp in certain location is more depending on the photoperiod, temperature, and precipitation during cultivation whereas the ecotype is less decisive.
Another study on different hemp cultivars and their productivity in certain climate conditions is the study of Sarsenbaev et al. [45]. This paper studied four different hemp cultivars and their productivity in the Moinkum desert. The seeds were planted in early May, and the hemp was harvested in the second half of September. The average rainfall in the valley falls of Chui which lay in the Moinkum desert is around 250 mm/year. Most of this rain falls in the middle of April. The maximum temperatures in the valley can reach up to 44 °C. The cultivars Fedora 17 and Lovrin 110 had higher stem lengths with irrigation whereas the cultivars Felina 34 and Futura 75 had higher stem lengths without irrigation. The variety Futura 75 also had a larger stem diameter in the absence of irrigation. The weight of stems with watering was higher for Felina 34 and Lovrin 110, and the weight of the stems of Fedora 17 and Futura 75 was heavier without watering. The overall productivity of stem kg/ha was higher for most of the cultivars with watering. Except for Futura 75, this cultivar's productivity was higher without watering when the plants were planted 30 cm apart. This paper concluded that the climate conditions (temperature and precipitation) in the Moinkum desert decreased fibre productivity of European hemp cultivars.

2. The sexual type of hemp can be divided in dioecious and monoecious. Monoecious cultivars contain both female and male flowers on the same plant, whereas in dioecious cultivars, male and female plants are separated [29]. The crop height and stem length are more homogeneous in monoecious cultivars than in dioecious cultivars [42]. Monoecious cultivars are better for dual harvest, but dioecious cultivars are preferable for fibre production [8]. Furthermore, male plants produce finer fibres and are better for high-quality textiles [48] but are more sensitive to pests [2]. There is a difference in fibre quality for each cultivar, but there are too many influencing factors, such as environment and the fibre extraction methods, to determine the exact fibre quality of each cultivar [2, 8].
3. The technical maturity of hemp is when the male plants are at full flowering, and when the female plants are at the peak of flowering [35]. Flowering is the main indicator of the correct harvesting time. When flowering starts, the nutrients of

the plants are used mainly for the growth of flowers and seeds instead of for the stems, leaves, and roots. The best time to harvest fibre hemp is right before the start of flowering when the stem yield is the highest [25, 50, 56]. During flowering, the production of secondary fibres increases, lignification intensifies, and the cellulose and pectin content decreases [30]. This all causes a decrease in primary bast fibres. The primary bast fibres are used for textile production since they are longer than the secondary fibres that are not suitable for textile end-use. Fibre quality is thus influenced by the maturity phase and the time of harvest [35]. Late flowering cultivars with a longer vegetative phase produce higher stem biomass and are more suitable for fibre production, whereas early cultivars are more suitable for seed production [36, 52]. Early to mid-early flowering cultivars can be used for dual harvest of fibres and seeds [10, 50].

The sexual type and flowering of hemp genotypes have an impact on the precipitation need of the cultivars. Early flowering monoecious fibre hemp needs around 250 mm of water, whereas late-flowering dioecious requires around 450 mm of water [14]. According to Cosentino et al. [14] monoecious cultivars flower on average from early to intermediate whereas dioecious cultivars flower on average from early too late. The sexual type also impacts the photoperiod. The critical photoperiod of male plants is longer than that of female plants of the same cultivar, as stated in the study of Borthwick and Scully [9]. Some other finds regarding monoecious and dioecious cultivars in the study of Petit et al. [40] were that monoecious cultivars have a better fibre quality than dioecious cultivars.

Some major breeding programmes in Europe are in France, Hungary, Poland, Romania, Italy, Spain, Russia, Germany, Czech Republic, and Slovakia [42]. The French breeding programme is mainly focused on variations that are cultivated for either pulp or fibres. This breeding programme aims at maintenance of the current cultivars and to reduce their THC content. The different variations of the high fibre cultivar 'Fibrimon', are named accordingly to when they are supposed to flower. The different Fibrimon cultivars are 'Fibrimon 21', 'Fibrimon 24', and 'Fibrimon 56'. The higher the number behind the cultivar's name, the later flowering [16]. The French cultivar Dioica 88 and the Italian cultivar Fibranova were developed to grow in Central-Southern environments and achieved higher biomass and dry stem yields [14]. The German-bred cultivar USO 11 is well adapted to grow in higher latitudes with long days, and the Polish-bred cultivar Beniko, Bialobrzeskie and the French cultivar USO 31 are also recommended for high latitudes [44].

China has been cultivating hemp for at least 6000 years [57]. There are currently around a hundred cultivars from China. The most well-known widely bred hemp variety in China are the YunMa cultivars [42]. The Yunma 2, Yunma 3, and Yunma 4 have a late maturity period and are dioecious. The female Yunma plants have thinner and longer fibres, and the male plants have shorter fibres with a larger diameter [33]. Some other hemp varieties that have been registered in certain areas but are not as widely used are the LongDaMa 1, JinMa 1, WangDaMa 1, and WangDaMa 2 cultivars [42]. JinMa 1 is grown in middle latitude regions in China and Yunma 7 in lower latitudes [59].

The cultivars that were bred in Canada can be used in various growing conditions. This breeding programme's main purpose is either for fibre, seed, dual-purpose hemp and plants that can be grown in specific environments. The Canadian breeding programme is not aimed at lowering the THC content, since the THC level in Canadian cultivars is well below the 0.3% allowed value [42]. The most common grown Canadian cultivars are Alyssa, Anka, CRS-1, CFX-1, CFX-2, Delores, and Finola [32].

The main differences between cultivars bred in these countries are the length of vegetation. In northern areas, fibre hemp cultivars with an early flowering period are used, whereas in southern regions, fibre hemp cultivars that flower late are used [42]. Early flowering cultivars are proposed for higher latitude regions, to avoid weather risks such as colder temperatures or high precipitation [28]. If early flowering cultivars are used in lower latitudes, the flowering will occur even earlier, which would result in lower biomass. Therefore, late-flowering cultivars are more suited for southern climates [42].

There are already many cultivars available, but the studies of Hall et al. [24] and Salentijn et al. [42] proposed improvements for future cultivars, such as a variety of fibre qualities, reduced lignin content, reduced amount of pectin, specific flowering times, and certain sexual types. Besides breeding programmes of hemp genotypes, hemp fibre yield can be affected by crop management [40]. Crop management practices include plant density and harvest date [40, 55].

3.2 Crop Management

Even though the impact of crop management, such as plant density and harvest date, on hemp fibres has been suggested in the literature [40], the conclusion from a study of Westerhuis [56] invalidated previous findings and literature since the effect of crop management on hemp fibres is only indirect. These indirect effects will be described more in detail below.

3.2.1 Plant Density

Plant density is measured as the number of plants per m^{-2}. Van der Werf et al. [54] studied the amount of dry biomass related to the density of surviving plants, which is also related to as self-thinning. Too high plant density results in dying crops and reduced plant growth. The highest stem yield is obtained at a planting density of 90 plants m^{-2}. Till this point, the stem quality improved, including bark content, whereas self-thinning decreases the amount of bark content in the stalks.

Westerhuis [56] also mentioned that a lower sowing density and late harvest resulted in an increased plant height and stem diameter. Next to that, certain plant height increases secondary fibre production. Secondary fibres cannot be used for textile production. As a part of crop management, it should be made sure that hemp crops stay below this height [56]. The higher the density, the earlier canopy will be, this causes an increase of humidity around stems and can result in fungal diseases [36].

3.2.2 *Harvest Date*

In general, dioecious hemp used for fibre production is normally harvested when the male crops finish flowering. The time of harvest depends on the latitude in which the crops are planted. In northern latitudes, this is usually around late August or early September, whereas in southern latitudes of the northern hemisphere harvest will be in July or early August [21].

Hemp harvest is mainly related to the flowering and maturing time of hemp. It is preferable to harvest hemp late but before flowering, since flowering decreases the stem growth. Furthermore, a delay in the harvest in certain regions increased the stem thickness; in other regions, the harvest date did not affect the stem diameter. Harvest after flowering decreases the number of high-quality fibres that are obtained. Hemp harvest after flowering reduces the stem part suitable for textile production [56].

The following subchapters will discuss the effects of the photoperiod and photosynthesis, temperature, precipitation, and soil on hemp growth in detail.

3.3 Photoperiod and Photosynthesis

The annual absorption of light depends on the crop emergence, speed of canopy formulation, and the plant's light interception. The light absorption can be impacted by the environmental factor's temperature, radiation, and daylength. Crop management can also influence the light interception through the choice of cultivar, plant density, sowing date, and harvest date [55]. Photoperiod mainly affects hemp in the second growth phase, the photosensitive phase [3].

Hemp is considered to be a short-day plant. Short-day plants require a dark period to activate the transformation phase from vegetative to the reproductive phase [24]. The critical photoperiod can be defined as the longest photoperiod a crop can allow, expressed in hours in a 24-h period [25]. The photoperiod regulates the flowering of hemp crops [24]. The average critical photoperiod of hemp is 12–14 h of daylight and 10–12 h of uninterrupted darkness [43]. However, it is also reported higher for French cultivars which have a critical photoperiod between 14 and 15.5 h [49].

The photoperiod influences the growth of hemp through the flowering period. The flowering period and growth period of hemp finish at the same time.

The relation between the photoperiod and maturing of hemp can be described as followed:

- An early flowering cultivar requires a shorter night, which results in a shorter growth phase, which eventually reduces stem growth.
- A later flowering cultivar needs a long night and has a longer growing phase, increasing the stem length.

If too many early flowering hemp plants are used, the fibre yield will suffer as a result of reduced stem growth [8]. Late-flowering cultivars accumulate higher yields and are therefore better for hemp fibre production [52].

If hemp plants are exposed to constant lightening for a 24-h regime, it will stop the plant's flowering and increase stem dry matter yield. Nevertheless, the stem's fibre content decreased [52]. Hemp flowers that are grown in a shorter photoperiod than the critical photoperiod will result in prematurely flowering. Crops that flower prematurely can be beneficial to limit the growing season of hemp crops [24]. A change in the photoperiod can also change the sex ratios in hemp cultivars and increases the possibility of monoecism crops [13]. Monoecious plants have more homogenous height and stem lengths [42] and are more suited for dual harvest for fibres and seeds, whereas dioecious plants are more suited for fibre end-use [8].

The photosynthesis of the canopy (and indirectly biomass production) and the interception of light are equivalent for non-stressed crops grown in favourable growing conditions [36]. The photoperiod is also related to radiation-use efficiency (RUE). RUE can be defined as the amount of dry matter yield compared to the amount of received photosynthetically active radiation (PAR) [37]. The RUE is dependent on the photosynthesis and respiration of the crops. RUE values of hemp are at a lower range than other C3 crops. A low RUE can be caused by low canopy photosynthesis, high respiration, or a large dry matter yield lost in the growing season [36].

Photoperiod and temperature are closely related when it comes down to the response of flowering time in a warmer environment. The flowering of crops is sensitive to both photoperiod and temperature will have a warmer optimum temperature rate, whereas the variability in flowering time decreases [15].

3.4 Temperature

The response of hemp, with regard to temperature, is measured by the ambient air temperature. Each plant has a minimum, maximum, and optimum temperature for growth [15]. The influence of temperature on hemp cultivation is the biggest in the juvenile phase (basic vegetation stage) [3]. The optimum temperature for photosynthesis and CO_2 uptake of C3 plants is between 20 and 30 °C [37]. The average temperatures that hemp is cultivated in are between 5.6 and 27.5 °C. The optimum temperature for hemp growth is at 14.3 °C [18]. The minimum temperature in which hemp can grow is −5 °C [6], and the maximum temperature is 40.7 °C [34].

Higher temperatures than the maximum temperature can accelerate the flowering period [15, 24]. A hastened flowering period reduces the overall yield and is therefore not desirable for fibre hemp [52]. Next to that, high temperatures can decrease the fibre quality [15, 24]. High temperatures and humidity can also increase the THC value in hemp yield [47]. Yield with THC values that surpass the maximum allowed percentage of the growing areas will be destroyed.

Hemp's growing stage is not dependent on colder periods [24], but freezing below the minimum temperature of hemp can decrease fibre yield [6]. Cold temperatures also prolong the flowering in hemp [24]. Longer flowering periods can increase the stem length [8]. The plants grown in relatively low temperatures of around 13 °C before or during flowering will likely get male flowers [9]. Hemp crops with male flowers produce finer fibres [48] but are more susceptible to pests [2].

Reduced yield due to high temperatures is often mentioned with precipitation rates and drought. Dry soils can be caused by an increase of evaporation due to high temperatures [22]. Drier soils also increase the intensity of droughts [51].

3.5 Precipitation

Precipitation, and the therewith associated water availability, is crucial for the development of crops. The results vary on the optimum amount of water required for hemp yield in the growing period, from sowing to harvest. Lisson and Mendham [34] estimated that the optimum amount is 535 mm, whereas Duke [18] mentioned 970 mm as the optimum water amount. According to Kraenzel et al. [31], hemp requires between 635 and 760 mm, and according to Bosca and Karus [10], it varies between 500 and 700 mm. The flowering and sexual type of cultivar can cause these water requirements differences [14].

Water stress has a negative impact on hemp yield and fibre quality. Studies identified reduced yields, plant growth, and fibre development due to water stress [1, 4, 12, 17]. In the study of Amaducci et al. [4], two hemp harvest years were compared. The first year was arid, with 96 mm rain from sowing to harvest. In the second year, the rainfall was closer to the optimum rain requirement for hemp growth. The stem height, stem yield, and fibre yield harvest were lower in the year with drier conditions. Abot et al. [1] reported similar outcomes, the stem height of the crop grown under water stress was 56% shorter than crops grown in non-water stressed environments. Next to that, the stem's diameter was smaller for crops that were cultivated in drier conditions. Another study, regarding the impacts of drought on bast fibre development, found that drought affects the bast fibre development and delays fibre elongation [12].

Besides the impact of water stress, heavy rain and hail also negatively impact hemp yield [11, 26, 38]. Torrential rains combined with heavy soils results in clogged grounds and eventually reduces hemp yield [11, 26]. Hail damages hemp crops which can lead to crop failure [38]. Long rain periods have also been linked to causing hemp diseases such as *Botrytis cinerea* and *Sclerotinia sclerotiorum* [36]. The cultivar Kompolti hybrid TC is more susceptible to these fungal diseases than other cultivars [53]. On the other hand, a lot of precipitation declines the amount of THC in hemp plants, which could be beneficial for cultivars close to the maximum allowed THC percentage [47].

3.6 Soil

Hemp can grow in various soils, but it reaches the highest yields when grown on well-drained loams with low acidity and high organic matter [17]. The optimum amount of pH value in the soil is 5.8–6.0 according to the study of Bosca and Karus [10] whereas the study of Amaducci et al. [2] indicated a higher amount of 6.0–7.5. Soil moisture should be between 50 and 70 cm in the vegetative growth for maximum yields [10]. Further in the growing process when the plant's roots are more developed, hemp can withstand drier soils. However, too much drought at the end of the growing season fastens the maturing of the crops and limits growth [17]. On the other hand, an abundance of soil moisture in poorly drained low-lying areas limits hemp growth and causes crop failure [17, 19].

Before sowing, soils should be ploughed and harrowed to create fine filth on top of the soil, which helps with weed suppression [56]. Poor soil structure furthermore can also result in waterlogging and soil compaction. This hinders growth and decreases crop homogeneously [17]. The soil structure can also impact the fibre quality, clay loams or heavier soils give coarser fibres than sandy loams or lighter soils [17].

3.7 Conclusion

Fibre hemp achieves its full potential when there is a sufficient amount of water, a good temperature, and enough light for optimal photosynthesis. The quality of hemp fibres is heavily impacted when one of these factors is limiting. However, the study of Petit et al. [40] concluded that the sensitivity of fibre hemp quality to the environment is dependent on the adaptation on the environment of the cultivars instead of the heritable genetic variation. Differences in temperature and critical photoperiod for particular cultivars are small, so common values can be used [3].

Hemp is considered as a well-adapted crop to varying climate conditions. However, the crop growth is also influenced by several agro-climatic factors such as the photoperiod, temperature, and precipitation. A short photoperiod, drought stress, and high temperatures can enhance early flowering. Early flowering caused by these factors decreases stem yield [3, 52]. Next to flowering, stem height, diameter, and fibre layers can be influenced by a lack of moisture [46]. At the same time, soil health and the amount of moisture in the soil impact yield and crop failure. Compact and crusted soil is prone to waterlogging and full saturation, which diminished hemp yield [11, 26].

A change in certain agro-climatic factors has also been proven to increase the THC value in hemp crops [47]. Sikora et al. [47] indicated the impact of temperature, humidity, and precipitation on THC percentage in hemp. Higher temperatures and humidity caused an increase in the THC value in hemp. High amounts of rainfall, on the other hand, decreased the THC value in crops. The psychoactive component THC is banned in many countries, and a maximum percentage of THC that is allowed in hemp is set in the producing countries. Therefore, an increase of THC could exceed this maximum permitted percentage which would result in a yield that needs to be destroyed. Another impact of temperature and precipitation vulnerability is also the development of hemp diseases. Higher precipitation has been associated with increased of hemp diseases [36].

References

1. Abot A et al (2013) Effects of cultural conditions on the hemp (*Cannabis sativa*) phloem fibres: biological development and mechanical properties. J Compos Mater 8(47):1067–1077. https://doi.org/10.1177/0021998313477669
2. Amaducci S et al (2015) Key cultivation techniques for hemp in Europe and China. Ind Crops Prod 68:2–16. https://doi.org/10.1016/j.indcrop.2014.06.041
3. Amaducci S, Colauzzi M, Bellocchi G, Venturi G (2008a) Modelling post-emergent hemp phenology (*Cannabis sativa* L.): theory and evaluation. Eur J Agron 28(2):90–102. https://doi.org/10.1016/j.eja.2007.05.006
4. Amaducci S, Zatta A, Pelatti F, Venturi G (2008b) Influence of agronomic factors on yield and quality of hemp (*Cannabis sativa* L.) fibre and implication for an innovative production system. Field Crops Res 107(2):161–169. https://doi.org/10.1016/j.fcr.2008.02.002
5. Are AK et al (2019) Chapter 3—application of plant breeding and genomics for improved sorghum and pearl millet grain nutritional quality. In: Taylor JRN, Duodu KG (eds) Sorghum and Millets, 2nd edn. AACC International Press, Eagan, pp 51–68. https://doi.org/10.1016/B978-0-12-811527-5.00003-4
6. BCMAF (1999) Industrial hemp (*Cannabis sativa* L.) factsheet. British Colombia Ministry of Agriculture and Food, Kamploops. Available at: https://www.votehemp.com/wp-content/uploads/2018/09/hempinfo.pdf. Accessed 10 Dec 2020
7. Bear R et al (2016) Principles of biology. New Prairie Press, Manhattan
8. Berenji J, Sikora V, Fournier G, Beherec O (2013) Genetics and selection of hemp. In: Bouloc P, Allegret S, Arnaud L (eds) Hemp industrial production and uses. CAB International, Oxfordshire, pp 48–71
9. Borthwick HA, Scully NJ (1954) Photoperiodic responses of hemp. Botanical Gazette 116:14–29. Available at: http://www.jstor.org/stable/2473219. Accessed 10 Dec 2020
10. Bosca I, Karus M (1998) The cultivation of hemp: botany, varieties, cultivation and harvesting. HEMPTECH, Sebastopol
11. Canadian Hemp Trade Alliance (2020) Impacts of severe weather events on hemp production. Available at: http://www.hemptrade.ca/eguide/production/impacts-of-severe-weather-events-on-hemp-production. Accessed 4 Dec 2020
12. Chemikosova SB, Pavlencheva NV, Gur'yanov OP, Gorshkova TA (2006) The effect of soil drought on the phloem fiber development in long-fiber flax. Russ J Plant Physiol 53(5):656–662. https://doi.org/10.1134/S1021443706050098

13. Clarke RC (1999) Botany of the genus cannabis. In: Ranalli P (ed) Advances in hemp research. Food Product Press an Imprint of the Haworth Press Inc., New York, pp 1–20
14. Cosentino SL et al (2013) Evaluation of European developed fibre hemp genotypes (*Cannabis sativa* L.) in semi-arid mediterranean environment. Ind Crops Prod 50:312–324. https://doi.org/10.1016/j.indcrop.2013.07.059
15. Craufurd PQ, Wheeler TR (2009) Climate change and the flowering time of annual crops. J Exp Bot 60(9):2529–2539. https://doi.org/10.1093/jxb/erp196
16. De Meijer E (1995) Fibre hemp cultivars: a survey of origin, ancestry, availability and brief agronomic characteristics. J Int Hemp Assoc 2(2):66–73. Available at: https://www.votehemp.com/wp-content/uploads/2018/09/jiha_vol2no2.pdf. Accessed 10 Dec 2020
17. Dewey LH (1914) The yearbook of the United States department of agriculture 1913. U.S. Department of Agriculture, Washington, D.C.
18. Duke JA (1982) Ecosystematic data on medical plants. In: Aktal CK, Kapur KM (eds) Utilization of medical plants. United Printing Press, New Dehli, pp 13–23
19. Ehrensing DT (1998) Feasibility of industrial hemp production in the United states pacific north west. Oregon State University, Oregon. Available at: https://www.votehemp.com/wp-content/uploads/2018/09/sb681.pdf. Accessed 7 Dec 2020
20. European Commission (2020) Agricultural species - Varieties. [Online] Available at: https://ec.europa.eu/food/plant/plant_propagation_material/plant_variety_catalogues_databases/search/public/index.cfm?event=SearchVariety&ctl_type=A&species_id=240&variety_name=&listed_in=0&show_current=on&show_deleted=. [Accessed 19 December 2020]
21. Fike J (2016) Industrial hemp: renewed opportunities for an ancient crop. Crit Rev Plant Sci 35(5–6):406–424. https://doi.org/10.1080/07352689.2016.1257842
22. Goosse H (2015) Climate system dynamics and modelling, 1st edn. Cambridge University Press, New York
23. Government of Canada (2020) List of approved cultivars for the 2020 growing season: industrial hemp varieties approved for commercial production. Available at: https://www.canada.ca/en/health-canada/services/drugs-medication/cannabis/producing-selling-hemp/commercial-licence/list-approved-cultivars-cannabis-sativa.html. Accessed 19 Dec 2020
24. Hall J, Bhattarai SP, Midmore DJ (2012) Review of flowering control in industrial hemp. J Nat Fibers 9(1):23–36. https://doi.org/10.1080/15440478.2012.651848
25. Hall J, Bhattarai SP, Midmore DJ (2014) The effects of photoperiod on phenological development and yields of industrial hemp. J Nat Fibers 11(1):87–106. https://doi.org/10.1080/15440478.2013.846840
26. Harper JK et al (2018) Industrial hemp production. The Pennsylvania State University, Pennsylvania. Available at: https://extension.psu.edu/industrial-hemp-production. Accessed 6 Dec 2020
27. Hillig K (2005) Genetic evidence for speciation in cannabis (Cannabaceae). Genet Resour Crop Evol 52:161–180. https://doi.org/10.1007/s10722-003-4452-y
28. Höppner F, Menge-Hartmann U (2007) Yield and quality of fibre and oil of fourteen hemp cultivars in Northern Germany at two harvest dates. Landbauforschung Völkenrode 57(3):219–232. Available at: https://literatur.thuenen.de/digbib_extern/bitv/dk038391.pdf. Accessed 17 Dec 2020
29. Johnson BL et al (2016) Industrial hemp cultivar evaluations in North Dakota. North Dakota State University, North Dakota
30. Keller A, Leupin M, Mediavilla V, Wintermantel E (2001) Influence of the growth stage of industrial hemp on chemical and physical properties of the fibres. Ind Crops Prod 13(1): 35–48. https://doi.org/10.1016/S0926-6690(00)00051-0
31. Kraenzel DG et al (1998) Industrial hemp as an alternative crop in North Dakota. The Insstitute for Natural Resources and Economic Development, North Dakota. Available at: https://www.votehemp.com/wp-content/uploads/2018/09/aer402.pdf. Accessed 18 Dec 2020
32. Laate EA (2012) Industrial hemp production in Canada. Government of Alberta, Alberta. Available at: https://www1.agric.gov.ab.ca/$department/deptdocs.nsf/all/econ9631/$file/Final%20-%20Industrial%20Hemp%20Production%20in%20Canada%20-%20June%2025%202012.pdf?OpenElement. Accessed 17 Dec 2020

33. Li X, Du G, Wang S, Meng Y (2015) Influence of gender on the mechanical and physical properties of hemp shiv fiber cell wall in dioecious hemp plant. Bioresources 10(2). https://doi.org/10.15376/biores.10.2.2281-2288
34. Lisson S, Mendham N (1998) Response of fiber hemp (*Cannabis sativa* L.) to varying irrigation regimes. J Int Hemp Assoc 1(5):9–15. Available at: http://www.internationalhempassociation.org/jiha/jiha5106.html. Accessed 11 Dec 2020
35. Mediavilla V, Leupin M, Keller A (2001) Influence of the growth stage of industrial hemp on the yield formation in relation to certain fibre quality traits. Ind Crops Prod 13(1):49–56. https://doi.org/10.1016/s0926-6690(00)00052-2
36. Meijer WJM, van der Werf HMG, Mathijssen EWJM, van den Brink PWM (1995) Constraints to dry matter production in fibre hemp (*Cannabis sativa* L.). Eur J Agron 4 (1):109–117. https://doi.org/10.1016/S1161-0301(14)80022-1
37. Moneith JL (1977) Climate and the efficiency of crop production in Britain. Philos Trans R Soc Lond 281:277–294. https://doi.org/10.1098/rstb.1977.0140
38. Pahkala K, Pahkala E, Syrjälä H (2008) Northern limits to fiber hemp production in Europe. J Ind Hemp 13(2):104–116. https://doi.org/10.1080/15377880802391084
39. Peel MC, Finlayson BL, Mcmahon TA (2007) Updated world map of the Köppen-Geiger climate classification. Hydrol Earth Syst Sci Discuss Eur Geosci Union 4(2):439–473. https://doi.org/10.5194/hess-11-1633-2007
40. Petit J et al (2020) Genetic variability of morphological, flowering, and biomass quality traits in hemp (*Cannabis sativa* L.). Front Plant Sci 11(102):1–17. https://doi.org/10.3389/fpls.2020.00102
41. Ranalli P (2004) Current status and future scenarios of hemp breeding. Euphytica 140:121–131. https://doi.org/10.1007/s10681-004-4760-0
42. Salentijn EMJ et al (2015) New developments in fiber hemp (*Cannabis sativa* L.) breeding. Ind Crops Prod 68:32–41. https://doi.org/10.1016/j.indcrop.2014.08.011
43. Salentijn EMJ, Petit J, Trindade LM (2019) The complex interactions between flowering behavior and fiber quality in hemp. Front Plant Sci 10(614). https://doi.org/10.3389/fpls.2019.00614
44. Sankari HS (2000) Comparison of bast fibre yield and mechanical fibres properties of hemp (*Cannabis sativa* L.) cultivars. Ind Crops Prod 11(1):73–84. https://doi.org/10.1016/S0926-6690(99)00038-2
45. Sarsenbaev K, Kozhamzharova L, Baytelieva A (2013) Influence high temperature, drought and long vegetation period on phenology and seed productivity European hemp cultivars in Moinkum Desert. World Appl Sci J 23(5):638–643. https://doi.org/10.5829/idosi.wasj.2013.23.05.13095
46. Schäfer T, Honermeier B (2006) Effect of sowing date and plant density on the cell morphology of hemp (*Cannabis sativa* L.). Ind Crops Prod 23(1):88–98. https://doi.org/10.1016/j.indcrop.2005.04.003
47. Sikora V, Berenji J, Latković D (2011) Influence of agroclimatic conditions on content of main cannabinoids in industrial hemp (*Cannabis sativa* L.). Genetika 43(3):449–456. https://doi.org/10.2298/GENSR1103449S
48. Small E (2015) Evolution and classification of cannabis sativa (marijuana, hemp) in relation to human utilization. Bot Rev 81:189–294. https://doi.org/10.1007/s12229-015-9157-3
49. Struik PC et al (2000) Agronomy of fibre hemp (*Cannabis satitiva* L.) in Europe. Ind Crops Prod 11(2–3):107–118. https://doi.org/10.1016/S0926-6690(99)00048-5
50. Tang K et al (2016) Comparing hemp (*Cannabis sativa* L.) cultivars for dual-purpose production under contrasting environments. Ind Crops Prod 87:33–44. https://doi.org/10.1016/j.indcrop.2016.04.026
51. Trenberth KE (2011) Changes in precipitation with climate change. CR Apecial 25, 47(1): 123–138. https://doi.org/10.3354/cr00953
52. Van der Werf HMG, Haasken HJ, Wijlhuizen M (1994) The effect of daylength on yield and quality of fibre hemp (*Cannabis sativa* L.). Eur J Agron 3(2):117–123. https://doi.org/10.1016/s1161-0301(14)80117-2

53. Van der Werf HMG, van Geel WCA, Wijlhuizen M (1995a) Agronomic research on hemp (*Cannabis sativa* L.) in the Netherlands, 1987–1993. J Int Hemp Assoc 2(1):14–17. Available at: http://druglibrary.net/olsen/HEMP/IHA/iha02107.html. Accessed 15 Dec 2020
54. Van der Werf HMG, Wijlhuizen M, De Schutter JAA (1995b) Plant density and self-thinning affect yield and quality of fibre hemp (*Cannabis sativa* L.). Field Crops Res 40:153–164. https://doi.org/10.1016/0378-4290(94)00103-J
55. Van der Werf HMG, Mathijssen EWJM, Haverkort AJ (1996) The potential of hemp (*Cannabis sativa* L.) for sustainable fibre production: a crop physiological appraisal. Assoc Appl Biol 129:109–123. https://doi.org/10.1111/j.1744-7348.1996.tb05736.x
56. Westerhuis W (2016) Hemp for textiles: plant size matters. Wageningen University, Wageningen. Available at: https://edepot.wur.nl/378698. Accessed 8 Dec 2020
57. Wulijarni-Soetjipto N, Subarnas A, Horsten S, Stutterheim N (1999) *Cannabis sativa* L. In: de Padua L, Bunyapraphatsara N, Lemmens R (eds) Plant resources of South-East Asia: no 12(1) medicinal and poisonous plants 1. Backhuys Publishers, Leiden, pp 167–175
58. Zatta A, Monti A, Venturi G (2012) Eighty years of studies on industrial hemp in the Po Valley (1930–2010). J Nat Fibres 9(3):180–196. https://doi.org/10.1080/15440478.2012.706439
59. Zhang Q et al (2018) Latitudinal adaptation and genetic insights into the origins of *Cannabis sativa* L. Front Plant Sci 9:1–13. https://doi.org/10.3389/fpls.2018.01876

Chapter 4
Climate Change Impact on Hemp

4.1 Climate Systems

To understand climate change, we need to understand the various climate systems. The climate system is a complex system which is influenced by a large number of fluxes and parameters. The four main climate systems are the atmosphere, the hydrosphere, the cryosphere, and the terrestrial biosphere [36].

The radiation balance in the atmosphere starts with an incoming flow of solar radiation, which is known as shortwave radiation. A part of this radiation is reflected by clouds or absorbed by the atmosphere. The remaining amount of radiation goes down to the surface, where another part is reflected and absorbed at the surface. Reflection at the surface happens for whiter objects with a high albedo. Albedo is the reflectiveness of an object regarding shortwave radiation. Objects with a high albedo are, for example, snow and ice, whereas asphalt has a low albedo. The absorbed solar radiation leaves the surface through evaporation, transpiration, sensible heat, and thermal up surface. Outgoing radiation from the earth surface is emitted as longwave radiation. The largest part of the emitted longwave radiation is absorbed by atmospheric greenhouse gases. Atmospheric greenhouse gases are gases that absorb longwave radiation and, in this way, affect the radiative properties of the atmosphere. A part of the absorbed longwave radiation is re-emitted back to earth. This re-emission significantly increases the temperature of the system [27].

The hydrological cycle has a large impact on the energy balance and contains out of, among others, the oceans, seas, rivers, lakes, and underground water. The hydrological gas, water vapour, is one of the most notable gases in the atmosphere. Water vapour turns into clouds when the invisible gases are turned into liquid water droplets. Clouds influence the energy balance through reflection of solar radiation and absorption and re-emission of longwave radiation. The water in the atmosphere is circulating quickly through precipitation, evaporation of oceans, and evapotranspiration of land and vegetation [27].

F. Dhondt and S. S. Muthu, *Hemp and Sustainability*, Sustainable Textiles: Production, Processing, Manufacturing & Chemistry,
https://doi.org/10.1007/978-981-16-3334-8_4

The cryosphere is the area on earth where water is in solid forms such as sea ice, lake and rive ice, snow cover, glaciers, ice caps, and ice sheets. Snow cover is the largest portion of solid water on earth. The cryosphere is very dependent on seasons. In the northern and southern hemisphere, snow and freshwater ice almost fully disappear in summer. Snow cover is the largest in the northern hemisphere since there is more land, whereas sea ice covers a similar area in the northern and southern hemisphere. The cryosphere influences the radiation fluxes through the albedo of snow and ice that reflects a large part of the incoming solar radiation. Snow and ice also play a role for global heat balance since it stores and releases latent heat. Furthermore, snow and ice also insulate the surface below it and decrease heat loss [27].

The terrestrial biosphere is the earth's soil, soil organic matter, and land vegetation [36]. The terrestrial biosphere is an important part of the climatic cycle for carbon storage and in particular in soils. The soil carbon sinks contain around triple the amount of carbon in the atmosphere and at least triple the amount that is stored in vegetation. Land-use changes can cause a change in the carbon stored in soils and result in more carbon in the atmosphere [31]. From 1850 onwards, 77% of the emissions from land-use and land-cover change were caused by deforestation [32]. Furthermore, plants take up carbon through photosynthesis [27]. The carbon uptake of the terrestrial biosphere varies per season [36]. Half of the uptaken carbon by plants is re-emitted, whereas the other half is stored in leaves, wood, and roots. Dead organic matter of plants is partly taken up in the soil carbon storage. The terrestrial biosphere is also connected to the hydrosphere since vegetated land holds more water than bare land [27].

4.2 Climate Change

The United Nations [55] defined climate change as a direct or indirect change of climate due to human activity that transforms the structure of the atmosphere above the average norms over a comparable time period. Climate change is not a new phenomenon as it has always occurred at earth. Paleoclimate studies showed that in the past, atmospheric carbon and climates also fluctuated [27]. However, what makes the latest climate changes remarkable is the pace in which it is happening. Enhanced amount of greenhouse gases released due to human activity heat the atmosphere through absorption and re-emission of radiation, eventually leading to increased temperatures. Burning of fossil fuels and deforestation are some of the major contributors. Burning of fossil fuels increases the amount of greenhouse gases in the atmosphere. These greenhouse gases act like a radiation blanket and absorb longwave radiation. A part of this is re-emitted back to earth, whereas another part is emitted to other atmospheric layers and eventually to space. Water vapour (H_2O (g)), carbon dioxide (CO_2), and methane (CH_4) are greenhouse gases with the largest impacts. These gases have always been present, but their presence has increased tremendously. Water vapour is the least dependent on human activities and mainly depends on oceanic temperatures. Carbon dioxide and methane have increased due to human activities, such as burning fossil fuels for carbon dioxide and agriculture and mining for methane. Land-use changes such as deforestation change the albedo in the area, which changes the amount of reflected solar radiation even further [31, 39].

The IPCC [39] reports on climate change impacts expects the global surface temperature to increase with 1.5 °C between 2030 and 2052 if emissions keep increasing at the same rate. The average of 1.5 °C fluctuates per region and season. Land regions are warming faster than oceans, and especially, in the polar regions, the temperature increases will be higher. This rise in temperature is also referred to as global warming. Global warming goes paired with rising seawater, extreme weather events, heat waves, drought, torrential rains, and biodiversity loss [39].

The IPCC predictions are currently based on climate models of four different representative concentration pathways (RCPs) regarding emission and concentration of greenhouse gases, air pollutant emissions, and land use [38]. Before IPCC introduced these RCPs in 2014, they used four storylines with six different special report emissions scenarios (SRES) of which the outcome is depending on the development of human society in terms of demographics and economic development, technological change, energy sources and demand, and land-use change [35]. Both the RCPs and SRES are addressed in this book since studies referred to use both pathways and storylines. These pathways are eventually expressed in solar radiation, temperature, precipitation, and soil moisture changes.

As indicated in Table 4.1 and Fig. 4.1, RCP 8.5 is comparable to the A2/A1F1 scenario, RCP 6.0 to B2, and RCP 4.5 to B1 [38]. RCP 2.6 is not similar to any of the SRES scenarios. The RCP pathways include a wider range of emissions than in the SRES scenarios [38].

It is hard to predict how climate change will affect hemp fibre production since there are multiple possible scenarios, and there is an uncertainty of the course of climate change. The scenarios fluctuate between more drastic or moderate changes and scenarios in between these two extremes. Simulated effects on geographical areas where hemp is cultivated can investigate climate change's possible impact [7]. The change in photosynthesis, temperature, precipitation, and soil composition determines the impact of climate change on hemp. All these factors are also bound to specific regions; hence, they will be further discussed in the next parts.

4.3 Photoperiod and Photosynthesis

Hemp crops respond to rising CO_2 in the atmosphere through photosynthesis, transpiration, and respiration [12, 22]. Climate change and human pollution will impact atmospheric aerosols, global dimming, and changes in atmospheric CO_2.

Photoperiod and photosynthesis can be impacted by an increase of aerosol in the atmosphere. The main components of the atmospheric aerosols are inorganic particles (such as sulphate, nitrate, ammonium, and sea salt), organic particles, black

Table 4.1 IPCC emission scenarios and pathways [35, 38]

IPCC RCP's [38]		IPCC SRES [35]
• **RCP 8.5** is the only scenario with increasing GHG emissions and for which it will be unlikely that temperature rise will stay below 3 °C in the twenty-first century	=	• **A2/A1F1** The A1 storyline exists out of three different scenarios based on their energy systems: A1F1 fossil intensive, A1T non-fossil energy sources or A1B a balance across all sources. A2 represents a heterogenic world with a focus on self-reliance and local traditions. The economic growth in this storyline is mainly regional. Economic development and technological change increase slower than in the other storylines
• **RCP 6.0** is a stabilisation scenario in which technologies and strategies reduce greenhouse gas emissions while avoiding an overshoot. The pathway will likely reach above 2 °C	=	• **B2** The B2 world focuses on local and regional levels with an orientation into environmental protection and social equity. The emphasis of this storyline is on local solutions to economic, social, and environmental sustainability. The model variables include population growth, economic development, generated power use, energy use efficiency, and mix of energy technologies, respectively
• **RCP 4.5** is a stabilisation scenario without exceeding the long-run target for climate forcing. The pathway will likely reach above 2 °C	=	• **B1** The B1 storyline portrays a convergent world with an economic structure mainly based on a service and information economy, including less materialism and the launch of clean technologies. This storyline emphasises worldwide solutions to economic, social, and environmental sustainability, which also includes enhanced equity
• **RCP 2.6** is the only pathway in which warming will stay below 2 °C and the only scenario with reduced GHG emissions		

carbon (fossil and biomass-based fuels), mineral species (mostly desert dust), and primary biological aerosol particles. Aerosols have a relatively short lifetime from a day up to ten days [37]. The amount of aerosol particles in the atmosphere that block sunlight is measured in the aerosol optical depth (AOD). A high level of AOD indicates many aerosol loadings in the atmosphere that pollute the atmosphere [6]. Aerosols are also divided in scattering and absorbing aerosols, scattering results in an overall cooling of the atmosphere, whereas absorption eventually leads to warming of the atmosphere. IPCC [37] also mentioned the possibilities of a regional

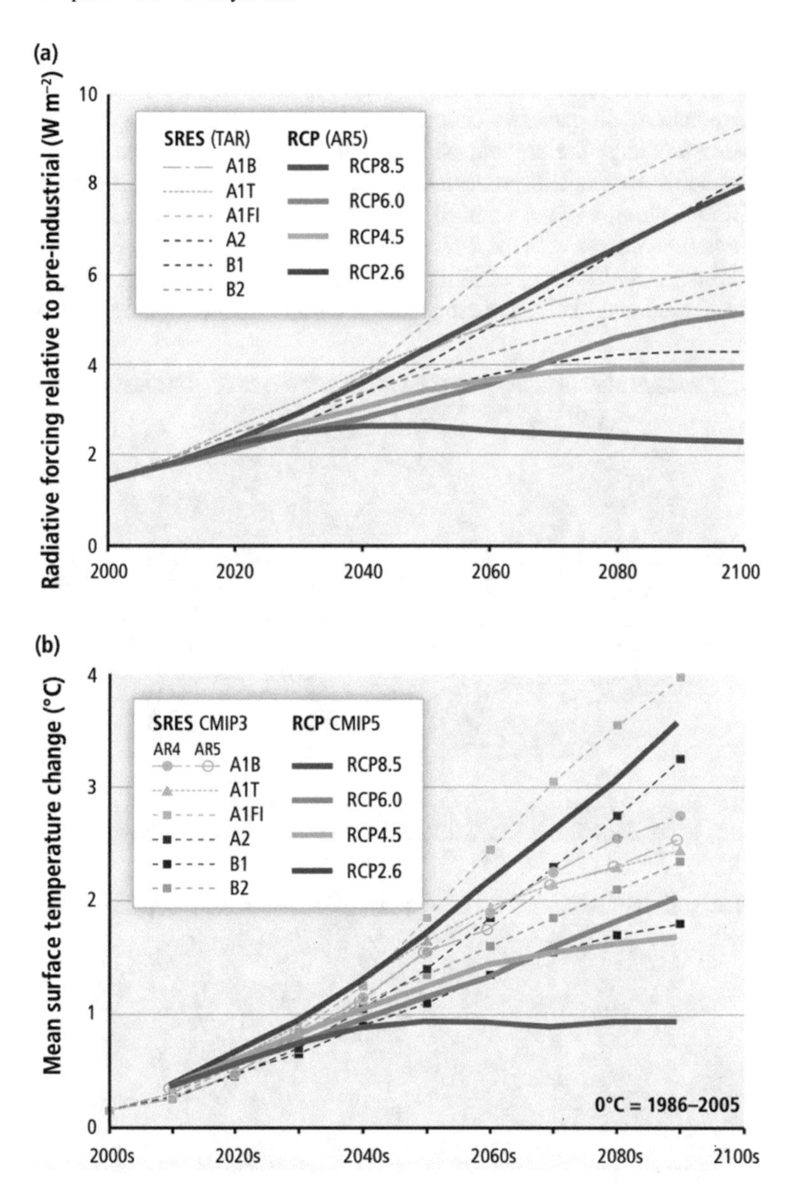

Fig. 4.1 **a** Projected radiative forcing (RF, W m^{-2}) and **b** global mean surface temperature change (°C) over the twenty-first century from the SRES and RCP scenarios [38].

change in precipitation patterns due to aerosols. The understanding of the impact of aerosols in the atmosphere and how these respond to climate change is limited [37].

Figure 4.2 shows the possible changes of solar radiation at the surface according to the RCP 8.5. The data is retrieved from the Royal Netherlands Meteorological Institute (KNMI) Climate Change Data and the World Meteorological Organisation

(WMO) [43]. As shown, the solar radiation in Europe, North America, and some parts of Africa and South America will increase, whereas in other parts of the globe, the solar radiation will decrease. Greenhouse gases, especially NO_2, H_2O (g), CH_4, and CO_2, can change the amount of incoming solar radiation. These greenhouse gases and aerosols result in processes referred to as global dimming and brightening. Global dimming is a reduced amount of incoming solar radiation at the surface, which increases with more clouds, greenhouse gases, and aerosols [36, 61]. Global dimming and aerosol changes that decrease direct radiation and increase diffused radiation will likely increase photosynthesis activities of crops [37, 46, 61].

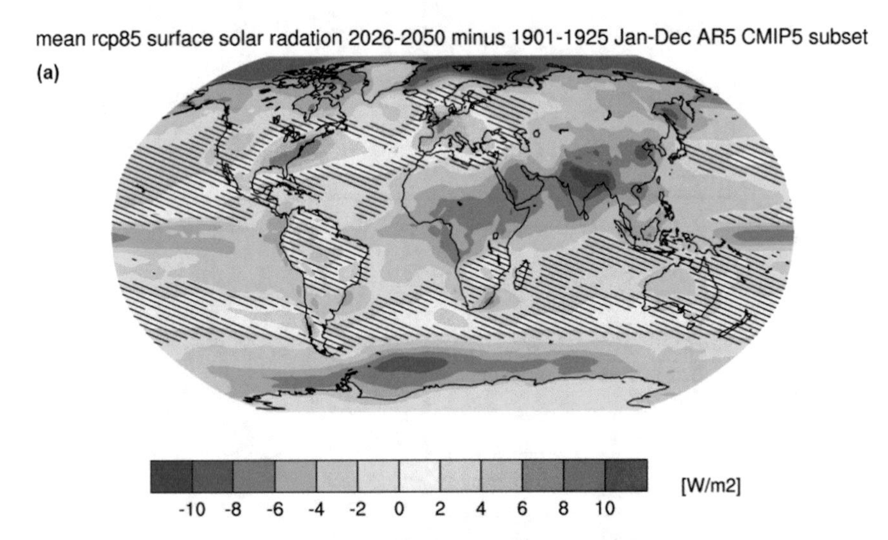

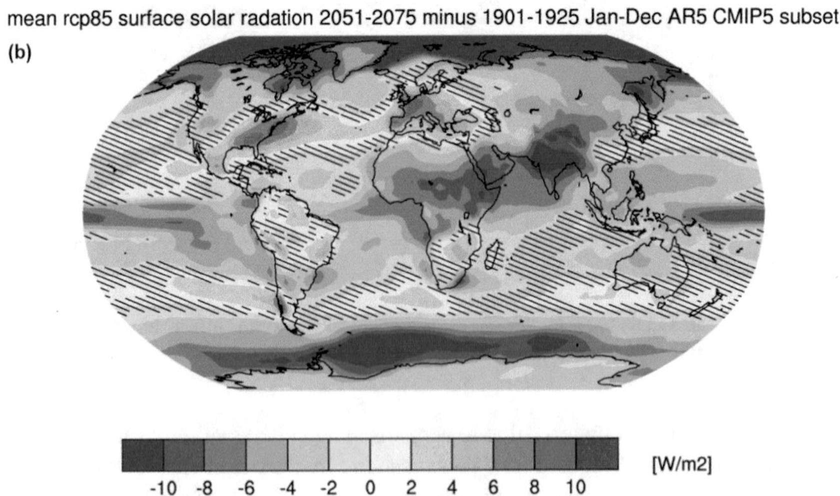

Fig. 4.2 **a** Surface solar radiation changes between 1901–1925 and 2026–2050 for IPCC RCP 8.5 **b** surface solar radiation changes between 1901–1925 and 2051–2075 for IPCC RCP 8.5 [43]

Besides the solar radiation that reaches the surface scattered through increased aerosols in the atmosphere, CO_2 also has an impact on photosynthesis. Increasing CO_2 concentrations have been linked to an increase of photosynthetic rates for C3 crops [17, 22, 42] and hemp crops [12]. The study of Drake et al. [22] reported that long-term exposure to atmospheric CO_2 increased the efficient use of nutrients, improved the soil–water balance, increased the uptake of carbon for crops grown in the shadow, and improved the carbon to nitrogen ratio for crops. The study of Chandra et al. [12] looked at the impact of increased CO_2 concentrations on hemp crops with regards to transpiration, leaf stomatal conductance, photosynthesis, water use efficiency, and leaf internal CO_2 concentration. Transpiration and leaf stomatal conductance decreased with higher CO_2 concentrations but photosynthesis, water use efficiency, and leaf stomatal conductance increased. This study concluded that hemp grown in higher CO_2 concentrations has the potential for improved growth and crop yield in dry but CO_2 rich climates.

4.4 Temperature

Hemp fibre yield is affected by the temperature in which it is cultivated. The optimum, minimum, and maximum temperatures for hemp to grow in are 14.3 °C [23], −5 °C [5], and 40.7 °C [45]. Temperatures above or below the maximum and minimum temperature can decrease the fibre yield and fibre quality [5, 16, 28].

Figure 4.3 shows the change in temperature according to RCP 8.5. In this scenario, the temperature will increase in every region of the globe. In northern regions and higher latitudes of the northern hemisphere, the temperature will change the most. Increasing temperatures can change the temperature in the growing area of hemp; as a result of this, the region might not suit the optimum temperature for hemp growth anymore. The growing areas of hemp can therefore shift to other regions that will suit the crop requirements better, in the future [57]. Even a small increase of temperatures that will cause temperatures to go above the optimum temperature can result in reduced productivity of the crop [59].

Furthermore, the updated Köppen climate classification of Fig. 4.4 is consulted to understand the shifts of climate regions with a temperature rise of 1.5 °C and above. The original Köppen climate classification was created by Wladimir Köppen to define climatic boundaries. The various climates were divided according to global temperatures and precipitation [50]. Rubel and Kottek [50] updated the Köppen-Geiger maps with the possible future special report emissions scenarios (SRES) estimated by IPCC [35]. Climate policies were not included in the SRES [38]. Figure 4.4 represents the scenario with the highest emissions and most noticeable change in climate [50].

Fibre hemp is most adapted to grow in the temperate climate zones [24]. The map of Fig. 4.4, which was created by Rubel and Kottek [50], displays a shift of the temperate climate zone. The temperate climate will move more towards the higher latitudes, especially in the northern hemisphere. Higher latitudes are from

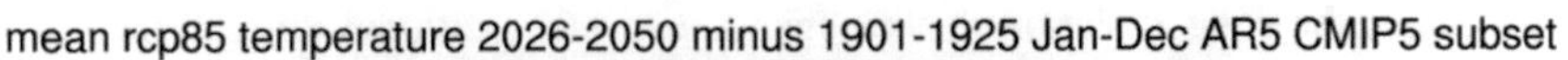

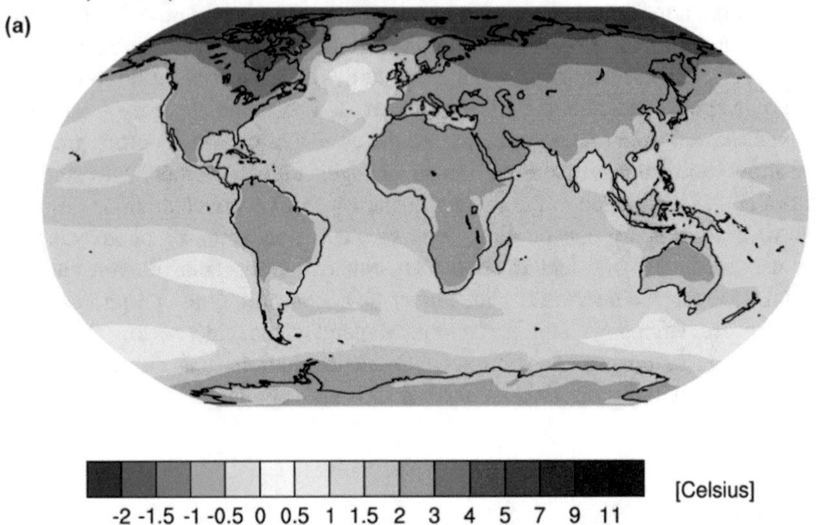

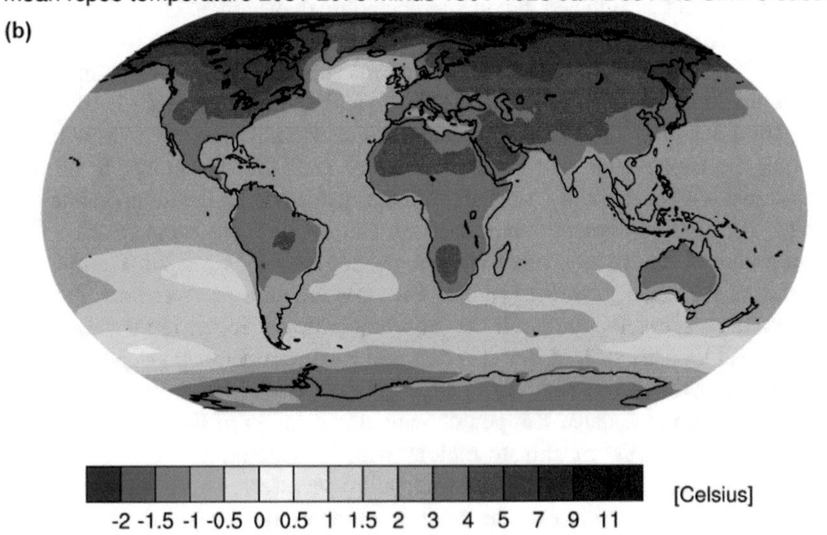

Fig. 4.3 **a** Temperature changes between 1901–1925 and 2026–2050 for IPCC RCP 8.5 and **b** temperature changes between 1901–1925 and 2051–2075 for IPCC RCP 8.5 [43]

60 degrees or up and −60 degrees or down from the equator. Farmers that grow hemp should already start taking into account that the climate in which they are currently growing can change. For hemp, it might not even be as bad, since the overall area that the temperate climate zone covers, will expand. Farmers who are

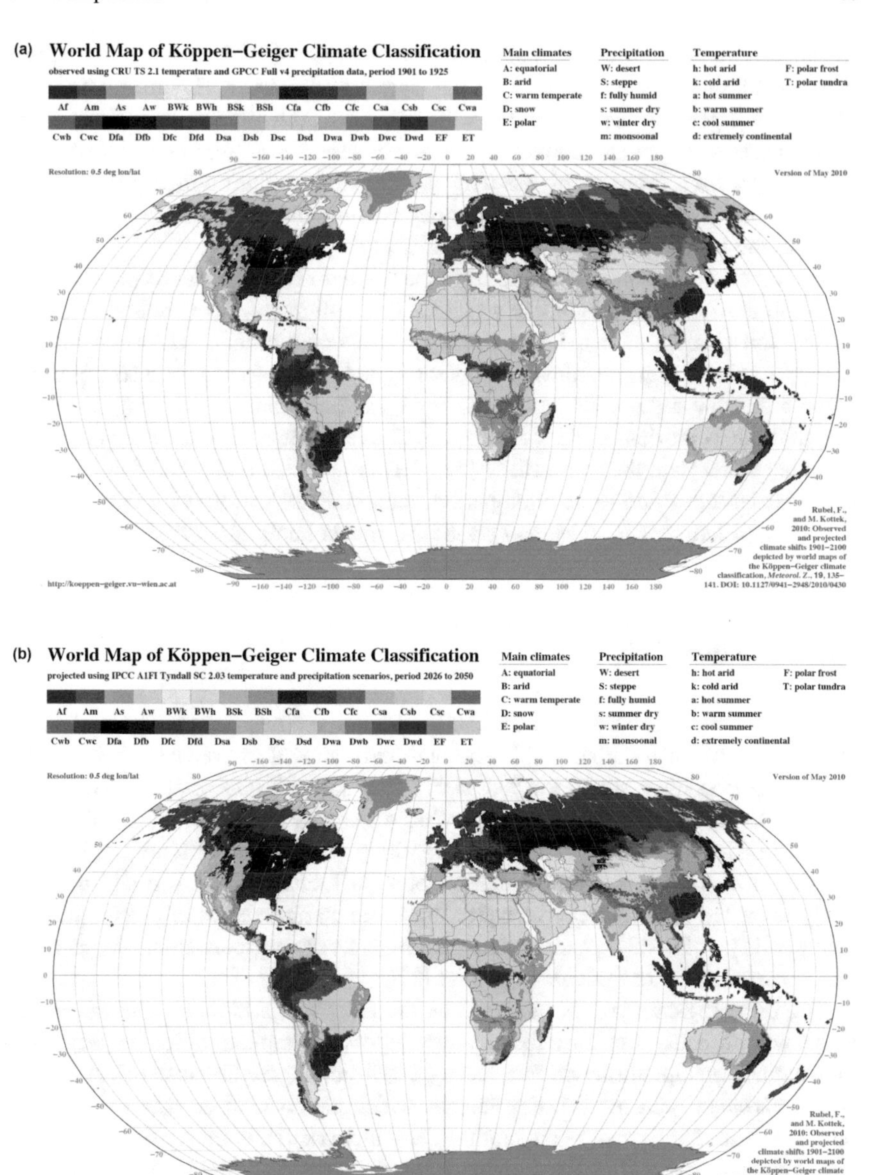

Fig. 4.4 **a** Köppen-Geiger climate classification 1901–1925 and **b** Köppen-Geiger climate classification 2026–2050 IPCC A1F1 scenario [50]

growing in Russia, Finland, and Norway will possibly encounter more favourable climate conditions for hemp growth [47]. Crop yield can possibly increase in higher latitudes due to climate change [19, 30, 53], and it might be possible to have multiple harvests per year [41]. Nevertheless, hemp that is grown in climates which are already at the top of the temperature boundaries will face more difficulties. Figure 4.4 shows that certain areas in the United States, India, Egypt, and Thailand [26, 47] will possibly be confronted with less favourable climate conditions as the temperatures will increase above the optimal boundaries of fibre hemp. Next to the limits of growing crops in higher temperatures than in the optimal growing temperature, increased temperatures cause drier growing environments for crops which increase the incidence of pests and diseases [40].

Even though the climate conditions in the northern latitudes can be more suitable for hemp in the future, it is unlikely that farmers will move. They will be held back by the soil quality, which limits the agriculture possibilities. The soils that are already fertile enough are already used for agricultural purpose. The soil fertilities in the areas that are not already in use for agricultures are not optimal, and it can take multiple centuries before it reaches the ultimate values. The climate is currently changing faster than the soil fertility [59]. Hemp will also be limited from growing in higher latitudes due to its photoperiod requirements. The varying day-length of the northern latitudes, higher than 64–65°N, affects hemp growth [8, 48].

4.5 Precipitation

Precipitation regulates the water availability for non-irrigated crops. The availability of water is crucial for fibre yield. The optimum amount of water for hemp in the growing period fluctuates from 500 to 970 mm according to scientists [9, 15, 23, 44, 45]. Especially, in the first six weeks, water availability is crucial for hemp growth. Older hemp plants can endure drier conditions, torrential rains, and short-term floods due to their extensive roots [56]. Hemp fibres can still be negatively impacted by both extreme water stress and an abundance of precipitation on heavy soils [11, 29, 48]. Precipitation has a higher variable than surface temperature and is therefore more difficult to predict [27].

The hydrological cycle is impacted by climate change [54]. Increasing temperatures will probably also raise the amount of global precipitation. It is expected that global mean precipitation will increase by 1–3% per degree Celsius [14]. A warmer atmosphere can hold more water vapour, which paired with an increase of evapotranspiration due to rising temperatures, and increase the intensity of precipitation [54]. Scientists expect that the precipitation will increase by 2100 in RCP 8.5, in the temperate moist-mid latitude zone. Hemp also grows in subtropical arid and semi-arid regions where precipitation will decrease, and farmers might encounter water shortage [14].

The change of precipitation for IPCC scenario RCP 8.5 is shown in Fig. 4.5. The change in mean precipitation is increasing in certain areas and decreasing in others, as can be seen in Fig. 4.5. In the higher latitudes of the northern hemisphere, rainfall will increase. Besides that, precipitation will also increase in the largest parts of Northern, Western, and Eastern Europe, India, the Middle East, and Central Africa. Precipitation will decrease for the countries that are located around the

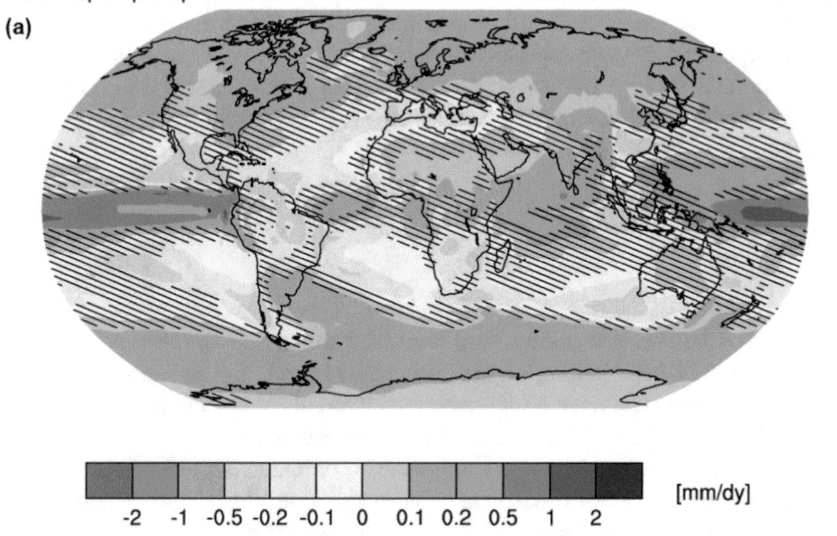

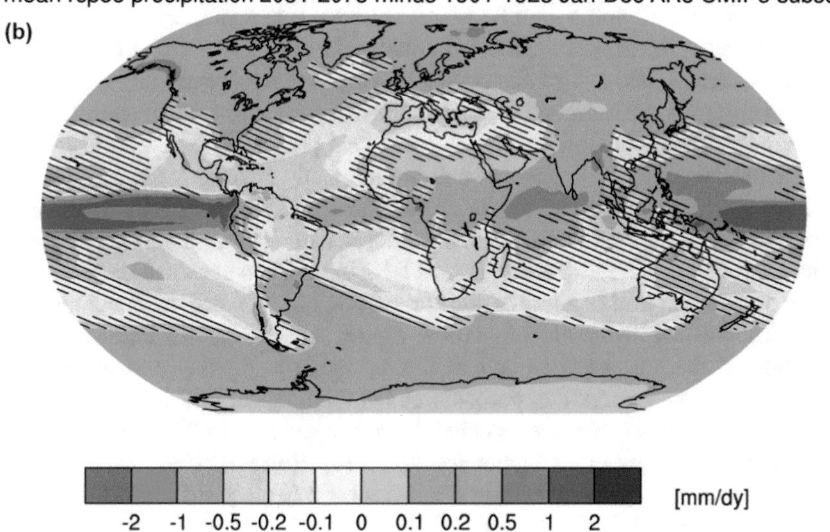

Fig. 4.5 Precipitation changes between **a** 1901–1925 and 2026–2050 for IPCC RCP 8.5 and **b** precipitation changes between 1901–1925 and 2051–2075 for IPCC RCP 8.5 [43]

Mediterranean Sea, Central America, South Africa, and certain parts of South America. Precipitation also decreases in the temperate zone in RCP 8.5, which contradicts the earlier findings of [14]. The yield of fibre hemp that is grown in these countries can be impacted due to these changes.

The amount of precipitation also influences the soil health and its hydraulic properties [33]. Precipitation and evapotranspiration have an impact on soil moisture and drought. Evapotranspiration is the movement of water in plants and soils that evaporates through the leaves and soil surface, while plants take up CO_2 and O_2. Precipitation rates that are higher than the evapotranspiration in a certain region will cause changes in soil moisture [14].

4.6 Soil

Hemp grows best on low acidity, well-drained loam soils [20]. Loam soils contain for 50% out of a combination of silt, clay, mainly sand, and 50% pore space and water [58]. Loam soils have a high saturated hydraulic conductivity that transport the water in the soils at a fast rate to deeper soil layers [34].

Figure 4.6 shows a change in soil moisture according to IPCC RCP 8.5. Most of the growing regions of hemp will face reduced soil moisture, whereas a few regions such as certain parts of China, India, Australia, and North America will have an increase in soil moisture. Temperature and precipitation impact the state of soil hydraulic properties. Soil hydraulic properties have a large impact on the productivity of crop growth [33]. Global warming directly influences soil hydraulic properties such as field water balance, soil moisture regime, water infiltration, water storage capacity, surface run-off, organic matter content, weathering, podsolization, and clay translocation. Besides that, the chemical properties such as pH, salinity, and the exchange organic matter of soil are indirectly affected by rising temperatures [33, 40]. In the earlier climate change time period, the moisture content in soils will be affected [40]. Rising temperatures can cause soil moisture content deficits through evapotranspiration [18, 51]. Rising temperatures and the correlated drying of soils can lead to a loss of soil carbon [40, 60]. Soil carbon is an important factor for the soil hydraulic properties. Losses of soil carbon influence other functions of the soil such as soil structure, stability, top layer water holding capacity, bulk density, texture, porosity, and soil biological functions [40].

Hemp has an extensive root system which can grow up to 200 cm and protect the plant from drought in the higher soil layers [3]. However, if extensive drought persists for a longer period, it can result in reduced hemp yield and a lower fibre quality [1, 2, 13, 20].

The texture of soil plays a big part in affecting different soil hydraulic properties. The structure of soils is heavily impacted by the amount of organic matter. A decrease in soil organic matter decreases the soil stability and infiltration and increases susceptibility to soil compaction, run-off, and erosion [10]. Soils with a finer texture, which hemp grows best in, have a higher sensitivity of organic carbon breakdown in warm temperatures than more coarse and sandy soils. Next to that, increasing temperatures enhances the vulnerability of soil organic matter in temperate climates over subtropical and tropical climates [21].

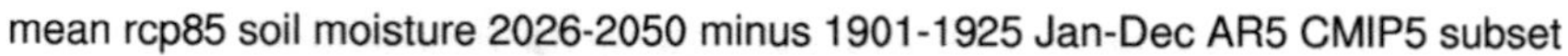

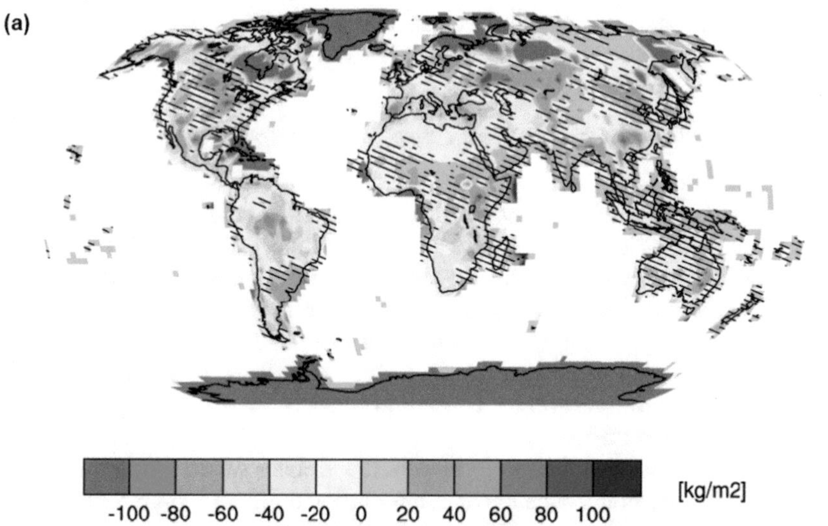

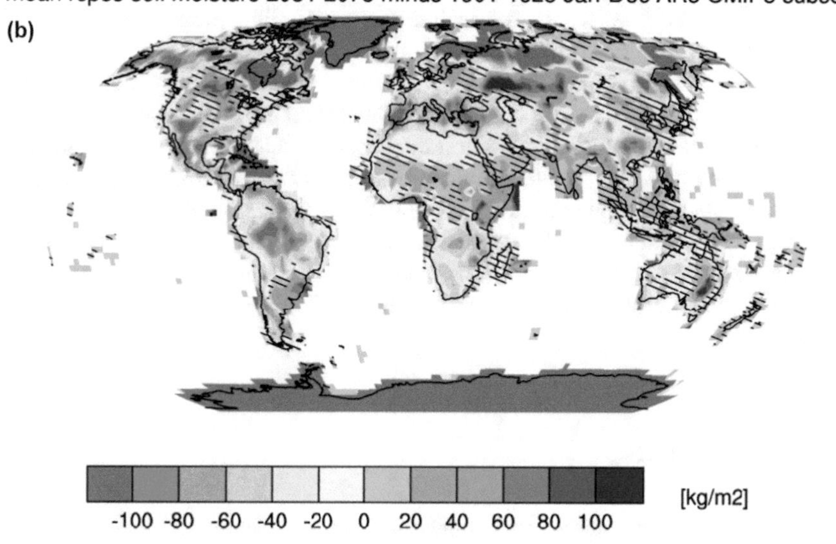

Fig. 4.6 **a** Soil moisture changes between 1901–1925 and 2026–2050 for IPCC RCP 8.5 and **b** soil moisture changes between 1901–1925 and 2051–2075 for IPCC RCP 8.5 [43]

In later climate change time periods, pH and soil acidity are affected as well [40]. Increased rainfall enhances leaching of basic cations, and this can cause more acidification. Biomass production, generated by higher temperatures and CO_2 in the atmosphere, also increases soil acidification. This will especially occur for crops that are grown and harvested for food or fibre since the biomass that would normally balance the pH value of the soil is removed from the field. CO_2 partials itself will likely not have an impact on soil pH [49].

4.7 The Unpredictability of Climate Change

Climate change is influenced by the human-induced emissions of the upcoming years. Therefore, it is uncertain what the exact future pathway will be. Besides understanding our own emissions, we also need to fully comprehend the climate systems for predictions to be correct. Some scientific risks which are not well understood are the thawing of permafrost, methane emissions, and other tipping points. It is important to understand these risks as well for future forecasts [52]. When studying future scenarios, it is better to consider what could happen, instead of what will happen [59].

Furthermore, the maps that were used in this chapter gave an overview of the global situation. Regional situation and impacts could differ of these results due to the large grid size of the maps [27]. Besides that, not every global map or model considers regional topography. Especially, precipitation is sensitive to topography [4]. Therefore, the final results for each region could differ when there is a more detailed local review of climate change impacts.

Uncertainty of climate change is important when we look at adaptation planning. Adaptation methods should be flexible to change with the changing climate. FAO [25] reported that the one of the biggest challenges is that climate change impacts, changes over time. The whole adaptation period including a change in climate change impacts should be considered when looking at adaptation methods. Another challenge is that arable land, and the economy of certain areas can be lost by extreme weather events. Climate is also bound to certain locations, impacts per location fluctuate heavily, and adaptations should therefore be focused on specific regions.

4.8 Conclusion

The possible future of hemp fibres is uncertain due to changing climate conditions and possible future scenarios. Some of the optimum climate conditions will move further up north, and some diminish, and other conditions can even improve hemp growth. An increased number of aerosols and increased CO_2 can improve the photosynthesis of C3 crops [17, 22, 37, 42, 46, 61], and hemp [12] which can be beneficial for hemp yield. The temperature maps showed an increase in global temperatures [43]. Temperature increase can have both a positive and negative impact on hemp growth. Regions that were considered too cold for hemp growth could become potential cultivation areas in the future. In contrast, the regions that are already used for hemp cultivation can exceed the maximum temperature for hemp growth. The Köppen-Geiger climate classification of the temperate zone [50] and the precipitation change showed a shift to northern regions for the optimal cultivation zones [43]. Soil moisture declined in most of the cultivation regions [43] and could reduce hemp growth and fibre quality in those areas [1, 2, 13, 20]. Hemp has some benefits over other crops [62] and is adaptable to multiple climates [24]; this might preserve it better from climate changes.

References

1. Abot A et al (2013) Effects of cultural conditions on the hemp (*Cannabis sativa*) phloem fibres: biological development and mechanical properties. J Compos Mater 8(47):1067–1077. https://doi.org/10.1177/0021998313477669
2. Amaducci S, Zatta A, Pelatti F, Venturi G (2008a) Influence of agronomic factors on yield and quality of hemp (*Cannabis sativa* L.) fibre and implication for an innovative production system. Field Crops Res 107(2):161–169. https://doi.org/10.1016/j.fcr.2008.02.002
3. Amaducci S, Zatta A, Raffanini M, Venturi G (2008b) Characterisation of hemp (*Cannabis sativa* L.) roots under different growing conditions. Plant Soil 313(1–2):227–235. https://doi.org/10.1007/s11104-008-9695-0
4. Basist A, Bell GD, Meentemeyer V (1994) Statistical relationships between topography and precipitation patterns. J Clim 7(9):1305–1315. https://doi.org/10.1175/1520-0442(1994)007%3c1305:srbtap%3e2.0.co;2
5. BCMAF (1999) Industrial hemp (*Cannabis sativa* L.) factsheet. British Colombia Ministry of Agriculture and Food, Kamploops. Available at: https://www.votehemp.com/wp-content/uploads/2018/09/hempinfo.pdf. Accessed 10 Dec 2020
6. Bellouin N et al (2011) Aerosol forcing in the climate model intercomparison project (CMIP5) simulations by HadGEM2-ES and the role of ammonium nitrate. J Geophys Res 116. https://doi.org/10.1029/2011JD016074
7. Bisbis MB, Gruda N, Blanke M (2018) Potential impacts of climate change on vegetable production and product quality—a review. J Cleaner Prod 170(1):1602–1620. https://doi.org/10.1016/j.jclepro.2017.09.224
8. Borthwick HA, Scully NJ (1954) Photoperiodic responses of hemp. Bot Gaz 116(1):14–29. https://doi.org/10.1086/335843
9. Bosca I, Karus M (1998) The cultivation of hemp: botany, varieties, cultivation and harvesting. HEMPTECH, Sebastopol
10. Bot A, Benites J (2005) The importance of soil organic matter key to drought-resistant soil and sustainaed food and production. FAO Soils Bulletin 80. Food and Agriculture Organization of the United Nations, Rome. Available at: http://www.fao.org/3/a0100e/a0100e00.htm. Accessed 10 Dec 2020
11. Canadian Hemp Trade Alliance (2020) Impacts of severe weather events on hemp production. Available at: http://www.hemptrade.ca/eguide/production/impacts-of-severe-weather-events-on-hemp-production. Accessed 4 Dec 2020
12. Chandra S, Lata H, Khan IA, Elsohly MA (2008) Photosynthetic response of *Cannabis sativa* L. to variations in photosynthetic photon flux densities, temperature and CO2 conditions. Physiol Mol Biol Plants 14(4):299–306. https://doi.org/10.1007/s12298-008-0027-x
13. Chemikosova SB, Pavlencheva NV, Gur'yanov OP, Gorshkova TA (2006) The effect of soil drought on the phloem fiber development in long-fiber flax. Russ J Plant Physiol 53(5):656–662. https://doi.org/10.1134/S1021443706050098

14. Collins M et al (2013) Long-term climate change: projections, commitments and irreversibility. In: Stocker T et al (eds) Climate change 2013: the physical science basis. Contribution of working group I to the fifth assessment report of the intergovernmental panel on climate change. Cambridge University Press, Cambridge, pp 1029–1136
15. Cosentino SL et al (2013) Evaluation of European developed fibre hemp genotypes (*Cannabis sativa* L.) in semi-arid Mediterranean environment. Ind Crops Prod 50:312–324. https://doi.org/10.1016/j.indcrop.2013.07.059
16. Craufurd PQ, Wheeler TR (2009) Climate change and the flowering time of annual crops. J Exp Bot 60(9):2529–2539. https://doi.org/10.1093/jxb/erp196
17. Cure JD, Acock B (1986) Crop responses to carbon dioxide doubling: a literature survey. Agric For Meteorol 38(1–3):127–145. https://doi.org/10.1016/0168-1923(86)90054-7
18. Dai A (2011) Drought under global warming: a review. WIREs Clim Change 2(1):45–65. https://doi.org/10.1002/wcc.81
19. Daliakopoulos IN et al (2017) Yield response of mediterranean rangelands under a changing climate. Land Degrad Dev 28(7):1962–1972. https://doi.org/10.1002/ldr.2717
20. Dewey LH (1914) The yearbook of the United States department of agriculture 1913. U.S. Department of Agriculture, Washington, D.C.
21. Ding F et al (2014) Decomposition of organic carbon in fine soil particles is likely more sensitive to warming than in coarse particles: an incubation study with temperate grassland and forest soils in northern China. PLoS ONE 9(4):1–10. https://doi.org/10.1371/journal.pone.0095348
22. Drake BG, Gonzalez-Meler MA, Long SP (1997) More efficient plants: a consequence of rising atmospheric CO2? Annu Rev Plant Physiol Plant Mol Biol 48:609–639. https://doi.org/10.1146/annurev.arplant.48.1.609
23. Duke JA (1982) Ecosystematic data on medical plants. In: Aktal CK, Kapur KM (eds) Utilization of medical plants. United Printing Press, New Dehli, pp 13–23
24. Ehrensing DT (1998) Feasibility of industrial hemp production in the United states pacific north west. Oregon State University, Oregon. Available at: https://www.ers.usda.gov/publications/pub-details/?pubid=41757. Accessed 10 Dec 2020
25. FAO (2008) Climate change adaptation and mitigation in the food and agricultural sector. Food and Agriculture Organization of the United Nations, Rome. Available at: http://www.fao.org/3/a-au034e.pdf. Accessed 10 Dec 2020
26. FAOSTAT (2020) Crops hemp tow waste. Available at: http://www.fao.org/faostat/en/#data/QC. Accessed 29 Dec 2020
27. Goosse H (2015) Climate system dynamics and modelling, 1st edn. Cambridge University Press, New York
28. Hall J, Bhattarai SP, Midmore DJ (2012) Review of flowering control in industrial hemp. J Nat Fibers 9(1):23–36. https://doi.org/10.1080/15440478.2012.651848
29. Harper JK et al (2018) Industrial hemp production. The Pennsylvania State University 2018, Pennsylvania. Available at: https://extension.psu.edu/industrial-hemp-production. Accessed 6 Dec 2020
30. He Q, Zhou G (2016) Climate-associated distribution of summer maize in China from 1961 to 2010. Agric Ecosyst Environ 232:326–335. https://doi.org/10.1016/j.agee.2016.08.020
31. Houghton J (2015) Global warming, the complete briefing, 5th edn. Cambridge University Press, Cambridge
32. Houghton RA, Nassikas AA (2017) Global and regional fluxes of carbon from land use and land cover change 1850–2015. Glob Biogeochem Cycles 31(3):456–472. https://doi.org/10.1002/2016GB005546
33. Indoria AK, Sharma KL, Reddy KS (2020) Hydraulic properties of soil under warming climate. In: Prasad MNV, Pietrzykowski M (eds) Climate change and soil interactions. Elsevier, Amsterdam, pp 473–508

34. Indoria AK, Sharma KL, Sammi Reddy K, Rao S (2017) Role of soil physical properties in soil health management and crop productivity in rainfed systems-I: soil physical constraints and scope. Curr Sci 112(12):2405–2414. https://doi.org/10.18520/cs%2Fv112%2Fi12%2F2405-2414
35. IPCC (2000) Special report on emissions scenarios. Cambridge University Press, Cambridge
36. IPCC (2007) Climate change 2007: the physical science basis. Contribution of working group I to the fourth assessment report of the intergovernmental panel on climate change. Cambridge Unitversity Press, Cambridge, United Kingdom and New York, United States
37. IPCC (2013) Climate change 2013: the physical science basis. Contribution of working group I to the fifth assessment report of the Intergovernmental Panel on Climate Change. Cambridge University Press, Cambridge
38. IPCC (2014) Climate change 2014: synthesis report. Contribution of working groups I, II and III to the fifth assessment report of the intergovernmental panel of climate change. In: Pachauri R, Meyer L (eds). IPCC AR5 synthesis report. IPCC, Geneva, p 151
39. IPCC (2019) Global warming of 1.5 °C. An IPCC special report on the impacts of global warming of 1.5 °C above pre-industrial levels and related global greenhouse gas emission pathways, in the context of strengthening the global response to the threat of climate change. World Meteorological Organization, Geneva
40. Karmakar R, Das I, Dutta D, Rakshit A (2016) Potential effects of climate change on soil properties: a review. Sci Int 4(2):51–73. https://doi.org/10.17311/sciintl.2016.51.73
41. Kawasaki K (2018) Two harvests are better than one: double cropping as a strategy for climate change adaptation. Am J Agric Econ 101(1):1–21. https://doi.org/10.1093/ajae/aay051
42. Kimball BA (1983) Carbon dioxide and agricultural yield: an assemblage and analysis of 430 prior observations. Agron J 75(5):779–788. https://doi.org/10.2134/agronj1983.00021962007500050014x
43. KNMI and WMO (2020) KNMI climate change atlas. Available at: https://climexp.knmi.nl/plot_atlas_form.py. Accessed 24 Jan 2021
44. Kraenzel DG et al (1998) Industrial hemp as an alternative crop in north dakota. The Institute for Natural Resources and Economic Development, North Dakota. Available at: https://www.votehemp.com/wp-content/uploads/2018/09/aer402.pdf. Accessed 18 Dec 2020
45. Lisson S, Mendham N (1998) Response of fiber hemp (*Cannabis sativa* L.) to varying irrigation regimes. J Int Hemp Assoc 1(5):9–15. Available at: http://www.internationalhempassociation.org/jiha/jiha5106.html. Accessed 11 Dec 2020
46. Mercado LM et al (2009) Impact of changes in diffuse radiation on the global land carbon sink. Nature 458:1014–1018. https://doi.org/10.1038/nature07949
47. Muzyczek M (2020) The use of flax and hemp for textile applications. In: Kozlowski RM, Mackiewicz-Talarczyk M (eds) Handbook of natural fibres volume 2: processing and applications. Taylor & Francis Group, Poznan, pp 147–168
48. Pahkala K, Pahkala E, Syrjälä H (2008) Northern limits to fiber hemp production in Europe. J Ind Hemp 13(2):104–116. https://doi.org/10.1080/15377880802391084
49. Rengel Z (2011) Soil pH, soil health and climate change. In: Singh BP, Cowie AL, Chan KY (eds) Soil health and climate change. Springer, Heidelberg, Berlin, pp 69–85
50. Rubel F, Kottek M (2010) Observed and projected climate shifts 1901–2100 depicted by world maps of the Köppen-Geiger climate classification. Meteorologische Zeitschrift 19(2):135–141. https://doi.org/10.1127/0941-2948/2010/0430. Publisher: http://www.borntraeger-cramer.de/journals/metz
51. Scheff J, Frierson DMW (2014) Scaling potential evapotranspiration with greenhouse warming. Am Meteorol Soc 27(4):1539–1558. https://doi.org/10.1175/JCLI-D-13-00233.1
52. Stern N (2016) Economics: current climate models are grossly misleading. Nat Comment 530:407–409. Available at: https://www-nature-com.proxy.library.uu.nl/news/economics-current-climate-models-are-grossly-misleading-1.19416. Accessed 11 Jan 2021
53. Supit I et al (2010) Recent changes in the climatic yield potential of various crops in Europe. Agric Syst 103(9):683–694. https://doi.org/10.1016/j.agsy.2010.08.009

54. Trenberth KE (2011) Changes in precipitation with climate change. CR Apecial 25 47(1): 123–138. https://doi.org/10.3354/cr00953
55. United Nations (1992) United Nations framework convention on climate change. United Nations, New York
56. USDA (1914) The yearbook of the United States department of agriculture 1913. U.S. Department of Agriculture, Washington, D.C.
57. USDA (2013) Climate change and agriculture in the United States: effects and adaptation. United States Department of Agriculture Technical Bulletin 1935, Washington, D.C. Available at: https://www.usda.gov/oce/climate_change/effects_2012/CC%20and%20Agriculture%20Report%20(02-04-2013)b.pdf. Accessed 11 Dec 2020
58. Vittum PJ (2009) Soil habitats. In: Resh VH, Cardé RT (eds) Encyclopedia of insects, 2nd edn. Elsevier, Burlington, pp 935–939
59. Wallace-Wells D (2019) The uninhabitable earth a story of the future, 1st edn. Penguin Books, London
60. Wan Y et al (2011) Modeling the impact of climate change on soil organic carbon stock in upland soils in the 21st century in China. Agric Ecosyst Environ 141(1–2):23–31. https://doi.org/10.1016/j.agee.2011.02.004
61. Wild M (2009) Global dimming and brightening: a review. J Geophys Res 114. https://doi.org/10.1029/2008JD011470
62. Zatta A, Monti A, Venturi G (2012) Eighty years of studies on industrial hemp in the Po Valley (1930–2010). J Nat Fibers 9(3):180–196. https://doi.org/10.1080/15440478.2012.706439

Chapter 5
Climate Adaptation

5.1 Adaptation

As the climate changes and environmental stress increases, it will affect yield and crop growth. To combat this, policymakers and farmers came up with solutions to adapt to the changing climate [22]. Policymakers are involved in climate change adaptation of certain regions. They are responsible for developing climate-resilient crops, regional adaptation plans, monitoring weather, warning of farms, coordinated planning, implementing innovations, technologies, and relocation of farm processing [31]. As stated by FAO [22], instead of performing in the same way, adaptation to climate change requires various measures that allow the agriculture sector to improve its production in changing environmental conditions.

An adaptation measure that is well known and already used on a larger scale is the greenhouse [56]. In a greenhouse, crops are grown inside where light, humidity, temperature, and water are entirely regulated [12, 56]. Yield in greenhouses is often higher than on the field, as the crop's optimum requirements can be more easily regulated. A downside of greenhouses is their environmental impact, as traditional greenhouses require lots of energy and freshwater compared to field farming.

Fortunately, more sustainable greenhouse systems are rising. There are already greenhouses that implement temperature regulation with solar thermal collectors, which eventually reduces fossil fuel use [56]. Nevertheless, even greenhouses might be affected by climate change, but the impact will be less than for crops grown in an open field [8]. Another more sustainable option that is less affected by the environment is hydroponic farms. In these farms, plants are not grown with soil but with mineral nutrient solutions in a water solvent [56]. These farms can be located anywhere, even underground [24, 56]. The possibility of farming underground is also an extra solution to the lack of arable land [7]. A downside of sustainable greenhouses and hydroponic farms is that the production costs are higher than conventional farming, making it less accessible for developing countries [56].

F. Dhondt and S. S. Muthu, *Hemp and Sustainability*, Sustainable Textiles: Production, Processing, Manufacturing & Chemistry,
https://doi.org/10.1007/978-981-16-3334-8_5

5.2 Temperature Management

Specific adaptation to increasing temperatures is needed as temperatures will likely increase by 1.5 °C between 2030 and 2052 [32]. Temperature management can be arranged through genotype engineering and on-farm adaptation measures. Furthermore, policymakers are responsible for monitoring and informing farmers on time when extreme temperatures are expected.

5.2.1 Policy Level

There are many different hemp genotypes. Some hemp fibre genotypes are more resistant to changing environments than others [48]. Farmers already search for hemp species that can withstand certain climate conditions [18]. Hemp farmers in New Mexico looked for alternative hemp strains as an adaptation method to abiotic stresses [26]. Besides using the already available genotypes, it is also possible to improve the current genotypes to withstand increasing temperatures better. The modulation of crop transcription factors is an effective method to increase environmental stress tolerance. Through this method, genetic engineering can be used in both traditional and new breeding technologies [4].

Some morphological traits which help with heath tolerance for crop breeding, according to Yadav et al. [58] are:

- Root length. Long roots are beneficial for crops as longer roots can take up more water and nutrients from the soil.
- Life span. Crops with a shorter lifespan are less prone to temperature effects as they are less exposed to heat stress in their shorter life cycle.
- Hairiness. Enhanced hairiness of crops creates more partial shade that repels sunlight from the cell wand and cell membrane.
- Leaf size. A smaller leaf size has a smaller stoma. Evaporation of the crop reduces due to the smaller stomata.
- Leaf orientation. A change in leaf orientation can increase the photosynthesis of the crop.
- Leaf glossiness and waxiness. More glossiness and waxiness of the leaves repels more sunlight and increases the heat tolerance of the crop.

Shuffling of genes may improve the thermal stability of RuBisCo. Photosynthesis, crop yield, and growth can increase in crops exposed to heat stress that is transformed through gene shuffling [37]. Another suggested option is lowering the photorespiratory flux. By lowering this flux, the inhibition of photosynthesis of C3 plants, caused by higher temperatures, reduces [35, 41]. The most familiar type of photosynthesis is C3 photosynthesis. C3 photosynthesis has to do with the type of carbon molecules produced during the process. C3 photosynthesis takes place in the mesophyll cells, which are located in the leaves. CO_2 is taken up and transported through the Calvin cycle, where it is fixed into 3-carbon molecules [6]. The photorespiratory bypass increases the temperature optimum for net photosynthesis in C3 crops [35, 41].

5.2.2 *Farm Level*

Besides genetic engineering, there are also other adaptation methods that farmers can use. As temperature increases due to climate change, the surface temperature will get closer to the maximum temperature of hemp. Possible adaptation methods to adapt hemp cultivation to raising temperatures are plant growth regulators and overhead netting.

Plant growth regulators are organic substances that control the growth development of crops. Environmental stresses affect crop growth and productivity. As a cause of ecological stresses, the endogenous hormones change. There are various plant molecules that play a large part in the thermotolerance of crops in heat-stressed environments. However, these mechanisms might fail to protect the crop against severe heat stress, resulting in plant death. The plant growth regulators create a defence system of the crop against high temperatures. The base of plant growth regulators is the chemical structure of these plant substances. In general, crops treated with plant growth regulators were more resilient to heat stress. The photosynthesis, leaf water status, and carbon allocation increase for treated crops.

The thermal protection of each crop responds differently. Therefore, more research should be done on how different crops respond to the crop growth regulation substances: phytohormones, osmoprotectants, and ROS scavengers [51].

The use of overhead netting or hail netting can reduce direct sunlight intensity and improve leaf-level photosynthetic light use efficiency [43]. Hail netting led to lower soil surface temperatures in the study of McCaskill et al. [42]. The netting also reduces wind speed compared to non-netted fields. However, if this wind speed is reduced by more than 25%, it can exacerbate temperatures. Therefore, the placement of the hail netting is essential. McCaskill et al. [42] left the field's sides open and only placed the net high above the crops. Hail netting also protects the crops from damage due to hail or birds [42].

5.3 Water Resource Management

Climate change will impact water availability in many regions due to changing temperatures, precipitation patterns, wind speed, vegetation cover, soil moisture, and runoff. Climate change can cause more floods, intensifying droughts, and changes in the long- and short-term water supplies [53]. Policymakers and farmers have a crucial role in adapting to the availability of water [31].

5.3.1 Policy Level

An approach on a policy level is needed to improve the resilience and adaptive capacity of different regions. The policy should include regional adaptation plans, monitoring and warning of farms, coordinated planning, and the implementation of innovations and technologies [31]. An invention that could be beneficial for areas in which freshwater is scarce is the desalination of seawater. The Sahara Forest Project already implemented this innovation in Jordan. A downside of desalination is that it requires a lot of energy. In desert areas with a lot of solar and wind energy available, it would be most beneficial since they can use green energy. Besides that, the project involves local communities but is very expensive to implement for smaller farms, making it a less available adaptation measure for some farmers [49]. Another way to innovatively use water is by the use of wastewater. Farmers should treat the wastewater before they can use it on their farmland. This method would mainly work for farms located close to urban areas [57]. Reclaimed wastewater can also benefit seed germination and crop yield due to the number of organic nutrients in the water [40, 46]. A combination of 50% wastewater and 50% distilled water resulted in higher seed germination results for hemp crops compared to 100% distilled water [38].

As the availability of water will change in certain regions, it is important that policymakers improve water use efficiency, reservoir capacity, water reuse, improve water charging and trade, re-negation of allocation agreements, set clear water use priorities, and integrate the demand for water in their systems [31]. Water use efficiency can be improved through water conservation with methods such as mulching, growing crops that are well adapted to the local climate conditions, proper pest and disease, fertility and time management, improved irrigation through either drip or deficit irrigation, and improving soil fertility [27].

As a response to floods, policymakers could change the rainfall interception capacity [31]. Rainfall interception is the amount of precipitation captured by the earth's surface and will eventually evaporate [23]. Increased rainfall interception reduces floods in areas that are vulnerable to it [31]. There are multiple types of rainfall interception, such as canopy, forest floor, fog, snow, and urban interception. Policymakers could influence urban interception, whereas farmers influence canopy interception [23]. For hemp, canopy rainfall interception is considered. Higher rainfall interception of hemp crops can be reached by increased planting density and a high leaf area index [19].

Policymakers can also respond to the increased irrigation request with improved climate change resilient crops [31]. The Global Initiative on Plant Breeding Capacity Build has been launched to enhance breeding programmes that focus on the genetic variability of crop variations. The initiative mainly concentrates on improving the breeding capacity in developing countries. The initiative is also supported by leading institutions, of which one of them is the Food and Agriculture Organization of the United Nations [20].

5.3.2 Farm Level

Farmers can respond to water availability changes with better soil moisture retention capacities, water storages, and water use efficiency on the farm [31]. One way to ensure better water use efficiency on farms is drip irrigation. Drip irrigation is a method in which low values of water is dripping on the soil from a plastic pipe with several outlets. The water is irrigated closer to the plant than in other irrigation methods [10]. Drip irrigation increases water use efficiency and shortens the growth period [22, 59] but does not affect the fibre content in hemp [47]. A detailed irrigation schedule is required for drip irrigation; otherwise, the crops can still suffer water stress. One of the downsides of drip irrigation is the costs of implementation. Therefore, the implementation of this depends on the farmers' social-economic situations [59]. Another irrigation method that improves water efficiency is deficit irrigation. With deficit irrigation, the water amount is kept below the evapotranspiration level. As more water is saved, farmers can irrigate more areas of land with the same amount of water. Deficit irrigation is especially beneficial for areas that are sensitive to drought since deficit irrigation might reduce yield for some crops [2].

Increased droughts also increase the need for irrigation. Farmers can respond to this by changing their crops or cropping patterns or improve agricultural practices that restore and retain soil moisture. A change in cropping patterns can ensure financial stability for farmers and reduce economic risks [31]. Farmers can change their cropping pattern to crop rotation. Crop rotation reduces evaporation and runoff but increases soil organic matter, soil moisture, and water infiltration [17]. Overall crop rotation can improve the soil quality and the number of nutrients in the soil [5]. Past research showed an increase in wheat yield when grown in rotation with hemp [3].

Furthermore, as a response to floods or extreme droughts, farmers can create or restore wetlands, including management for floods, improve the drainage systems and use crops resistant to droughts [31]. Climate change will worsen the number of floods and droughts in the future [28]. Wetlands are a suitable adaptation method against both droughts and floods. Wetlands are an area of land or soil covered with a layer of water and are high in biodiversity. Wetlands have many beneficial properties and can also be used as an adaptation method against high temperatures and heat waves [25]. Wetlands can be used in-between growing periods for hemp but should be avoided during hemp growth as hemp growth is limited on wet soils [18, 54].

Another adaptation possibility against floods and waterlogging is the use of agricultural drainage systems. These systems reduce the water level on the field. There are two types of systems, where the water can either flow through the ground through a subsurface drainage pipe or over the surface [45, 55]. A drainage system can also control salinisation and erosion [55]. Well-drained soils improve hemp yield [54].

Large strategic decisions are dependent on governance, the participation of farmers, and information and communication technologies [11].

5.4 Soil Health

Soil health is an essential factor for hemp growth [18]. Therefore, the health of the soil changes due to changing climate circumstances is of great importance for hemp fibre yield. Organic matter improves soil health and can be an adaptation method against drought and heavy rain [20]. Some other examples proposed by FAO [20, 21] are cover cropping and mulch cropping. Another problem caused by climate change is salinisation. Salinisation decreases hemp growth and should, therefore, be avoided [29].

5.4.1 Organic Matter

Hemp grows best on soils with a high organic matter [18]. Organic matter in the soil is a large variety of microorganisms, including plants, insects, and animals. Soil organic matter improves soil structure and increases the amount of water that the soil can absorb. Therefore, soil organic matter is a crucial adaptation method to respond to drought and heavy rain [20]. Organic matter is also important for crop yield as it influences the availability of nutrients and influences fertilisers' effect [9].

The amount of organic matter in the soil can be influenced by environmental stresses [58]. Soil organic matter increases with lower temperatures, increased annual precipitation, clay content, salinity, and extreme pH values in the soil. Soil organic matter is also influenced by topography. The organic matter is higher at the foot of a mountain or hill [9]. Agricultural practices that can influence soil organic matter are cover crops, crop rotation, drainage, and tillage of the soil [9, 13, 17]. Cover crops and crop rotation increase the amount of soil organic matter [9, 13, 17], whereas tillage and high drainage reduce the amount of organic matter in the soil [9].

5.4.2 Cover Cropping

Cover crops cover the soil and protect the soil from nutrient loss and soil erosion [50]. There are multiple reasons why cover crops can be beneficial for soil health and crop productivity. Some benefits of cover crops are prevention of nitrogen leaching, improved soil structure, enrichment of soils by nitrogen fixation, and control of soil-borne illnesses [50]. Other advantages of cover crops are suppression of weeds, blocking light, and alteration of incoming light waves that eventually change soil surface temperature, reduces soil diseases through better microbial life. Cover crops also reduce the number of herbicides and pesticides that are needed. Cover crops increase water infiltration speed, add organic matter to the soil, and prevent soil erosion. Cover crops eventually improve the yields of cash crop by improving soil health [13]. The crops are not grown to yield economic benefits but rather enhance soil quality [16]. Cover crops are useful for climate change adaptation due to their ability to reduce soil vulnerability to erosion caused by torrential rain, improve soil water management in case of drought or soil saturation and conservation of mineralising nitrogen due to increasing temperatures [34].

There are several categories of cover crops. Winter cover crops are often either grasses, legumes, or brassicas. Legume cover crops have the ability to transform atmospheric nitrogen into nutritional nitrogen, which is beneficial for soil health and can replace the need for fertilisers [33]. The paper of Jones et al. [33] determined the following cover crops as a potential for hemp: cereal rye, oats, wheat, barley, triticale, annual ryegrass, hairy vetch, field peas, clovers, and brassicas. When implementing cover crops for hemp fibre use, it is important to consider the seeding rate, timing, requirements for nutrients, and crop management [33].

Cover crops can be implemented in multiple ways. It can be grown in-between the growing seasons of the summer cash crops. In this case, the cover crop will grow, most of the time, in fall and winter. It is also possible that the cash crops grow in the winter, fall, and spring periods, then the cover crop will grow in summer. Another method is the overlapping of the cover crops' growth period with the last growth period of the cash crop. If a cash crop is not sown annually, the cover crop can grow for multiple years until the cash crop is grown again. The last method is living mulch; for this management option, the cover crop is grown in-between the cash crops' rows [50].

5.4.3 Mulch Cropping

Another adaptation method is mulch cropping. Mulch can either be inorganic or organic. Inorganic mulch covers the soil with a plastic firm, whereas organic mulch protects the soil with crop residue. Mulch coverings are often used to moderate soil temperature, erosion, enhance soil fertility, structure, and minimise weed growth. Mulching is a relatively accessible and affordable technology for farmers [58].

For hemp fibre production, the residues of the leaves can be used for mulch cropping. The hemp mulch can improve soil moisture and increase the soil's microbial biodiversity [36]. Next to that, hemp that is retted on the field leaves behind residual nutrients that can be beneficial for crops grown the year after it on the same field [1].

Mulch combined with crops also delays soil runoff which leads to higher water absorption of the soil. Furthermore, mulch cropping reduces the amount of water that evaporates from the soil and increases the amount of water that infiltrates the soil. More water can infiltrate into the soil due to the roots and organic matter that increase the soil's pores and channels. The combination of reduced evaporation and a higher water infiltration limit the amount of salt that moves upward [14].

5.4.4 Salinisation

Salinisation is another climate change problem that can impact the growth of hemp [29]. Salinisation in the soil can delay the start of germination and represses the length of the embryonic root and stem of hemp crops [30]. An increased concentration of salt also resulted in a decrease in plant height, dry weight, and root length [29].

Salinisation is the build-up of water-dissolved salts in the upper part of the soil that exceeds a certain value [44]. Salt limits the ability of the crop to take up water. Soil salinity is a natural process, whereas soil salinisation is human-induced due to poor water and land management. Soil salinisation is the increased amount of salt in the soil, which makes it harmful to crops. Soil salinity is caused by rock weathering, volcanic activity, deposition in coastal areas, and wind. Human-induced salinisation is caused by overirrigation, irrigation with saline groundwater, change in evapotranspiration [60], rising sea levels, and low precipitation. Water stressed regions in South and Southeast Asia are especially vulnerable to soil and groundwater salinisation [39, 52]. Climate change will intensify salinisation due to the rising sea levels, which will intensify the amount of saltwater that will reach coastal areas and increased evapotranspiration due to rising temperatures [44].

One of the adaptation methods for soil salinisation is to leach the soil with low-salt, high-quality water. It requires flushing, irrigation, and drainage to lower the salt levels in the soil. The effectiveness of flushing is dependent on some chemical and biological processes. As some salts dissolve more easily than others, salts that barely dissolve require more water for flushing. Furthermore, a competent drainage system is required to lower the groundwater and drains the saltwater off [44]. However, this option is too expensive for various farmers. To lower costs, it is also possible to mix desalinated water and well water [15]. Improved fertilisation management can also be used as an adaptation method. A limitation of fertilisers that include salts will decrease salinisation [39]. Another possible adaptation for hemp crops is the use of different cultivars. As an adaptation method to salinisation, it is possible to use the Chinese hemp cultivar YM5. YM5 showed a higher salt tolerance than the other studied hemp cultivar BM [29, 30].

5.5 Conclusion

Adaptation is needed as temperatures will likely rise between 2030 and 2052, the amount of available water will be uncertain, and soil health will be impacted [32]. Changing climate conditions requires crop adaptation measures on both policy and farm levels [31]. On a policy level, the development of heat tolerance genotypes is of huge importance to ensure future hemp yield. Farm measures such as greenhouse farming, hydroponic farming, plant growth regulators, and overhead netting can protect crops from increasing temperatures up to a certain point. Water availability requires even more policy measures that improve water use efficiency, reservoir capacity, water utilisation, improve water charging and trade, re-negation of allocation agreements, set clear water use priorities, and integrate the demand for water in their systems [31].

Furthermore, innovations such as the desalinisation of seawater might be able to ensure future irrigation needs. Farmers are also responsible for efficient water use and the continuity of soil health. Both policy and farm levels are important for adaptation methods to ensure future crop yields and the adaptative capacity of these crops. There are already innovations and adaptation methods that can be implemented as future adaptation methods in developed countries. Most of these innovations are not accessible for developing countries as the production costs would rise too much. More accessible innovations are needed to ensure crop yields in developing countries.

References

1. Adesina I, Bhowmik A, Sharma H, Shahbazi A (2020) A review on the current state of knowledge of growing conditions, agronomic soil health practices and utilities of hemp in the United States. Agriculture 10(129):1–15. https://doi.org/10.3390/agriculture10040129
2. Ali MH, Hoque MR, Hassan AA, Khair A (2007) Effects of deficit irrigation on yield, water productivity, and economic returns of wheat. Agric Water Manag 92(3):151–161. https://doi.org/10.1016/j.agwat.2007.05.010
3. Altarriba GG, Vilamanyà JL (2000) Effect of hemp (*Cannabis sativa* L.) in a crop rotation hemp–wheat in the humid cool areas of North-Eastern of Spain. In: Parente G, Frame J (eds) Crop development for the cool and wet regions of Europe. European Comission, Brussels, pp 99–106
4. Anderson R, Bayer PE, Edwards D (2020) Climate change and the need for agricultural adaptation. Curr Opin Plant Biol 56:197–202. https://doi.org/10.1016/j.pbi.2019.12.006
5. Aziz I, Ashraf M, Mahmood T, Islam KR (2011) Crop rotation impact on soil quality. Pak J Bot 43(2):949–960. Available at: http://www.pakbs.org/pjbot/PDFs/43(2)/PJB43(2)0949.pdf. Accessed 10 Feb 2021
6. Bear R et al (2016) Principles of biology. New Prairie Press, Manhattan
7. Benke K, Tomkins B (2017) Future food-production systems: vertical farming and controlled environment agriculture. Sustain Sci Pract Policy 13(1):13–26. https://doi.org/10.1080/15487733.2017.1394054
8. Bisbis MB, Gruda N, Blanke M (2018) Potential impacts of climate change on vegetable production and product quality—a review. J Cleaner Prod 170(1):1602–1620. https://doi.org/10.1016/j.jclepro.2017.09.224
9. Bot A, Benites J (2005) The importance of soil organic matter. Key to drought-resistant soil and sustained food production. Food and Agricultulture Organization (FAO), Rome. Availabe at: http://www.fao.org/3/a0100e/a0100e00.htm#Contents. Accessed 30 Jan 2021
10. Brouwer C, Prins K, Kay M, Heilbloem M (1988) Irrigation water management: irrigation methods. Food and Agriculture Organization (FAO), Rome. Available at: http://www.fao.org/3/s8684e/s8684e00.htm#Contents. Accessed 29 Jan 2021
11. Cameira, MdR, Santos Pereira L (2019) Innovation issues in water, agriculture and food. Water 11(6):1230. https://doi.org/10.3390/w11061230

12. Cantieri J (2018) Innovative greenhouses help farmers adapt to climate change. Available at: https://www.nationalgeographic.com/environment/future-of-food/telangana-india-agriculture-greenhouses/. Accessed 2 Feb 2021
13. Clark A (2008) Managing cover crops profitably, 2nd edn. Sustainable Agriculture Network, Beltsville
14. Corsi S, Muminjanov H (2019) Conservation agriculture. Training guide for extension agents and farmers in Eastern Europe and Central Asia. Food and Agriculture Organization (FAO), Rome. Available at: http://www.fao.org/3/i7154en/i7154en.pdf. Accessed 26 Jan 2021
15. Cuevas J et al (2019) A review of soil-improving cropping systems for soil salinization. Agronomy 9(295):1–22. https://doi.org/10.3390/agronomy9060295
16. Daryanto S et al (2018) Quantitative synthesis on the ecosystem services of cover crops. Earth Sci Rev 185:357–373. https://doi.org/10.1016/j.earscirev.2018.06.013
17. de Azevedo DMP, Landivar J, Vieira RM, Moseley D (1999) The effect of cover crop and crop rotation on soil water storage and on sorghum yield. Pesq Agrop Brasileira 34(3):391–398. https://doi.org/10.1590/s0100-204x1999000300010
18. Dewey LH (1914) The yearbook of the United States deparment of agriculture 1913. U.S. Department of Agriculture, Washington, D.C.
19. Drastig K, Flemming I, Gusovius H-J, Herppich WB (2020) Study of water productivity of industrial hemp under hot and dry conditions in Brandenburg (Germany) in the year 2018. Water 12(11):2982. https://doi.org/10.3390/w12112982
20. FAO (2007) Adaptation to climate change in agriculture, forestry and fisheries: perspective, frameworks and priorities. Food and Agriculture Organization (FAO), Rome. Available at: http://www.fao.org/3/a-au030e.pdf. Accessed 5 Feb 2021
21. FAO (2008) Climate change adaptation and mitigation in the food and agricultural sector. Food and Agriculture Organization (FAO), Rome. Available at: http://www.fao.org/3/a-au034e.pdf. Accessed 10 Feb 2021
22. FAO (2013) Climate-smart agriculture sourcebook. Food and Agriculture Organization (FAO), Rome. Available at: http://www.fao.org/3/i3325e/i3325e.pdf. Accessed 3 Feb 2021
23. Gerrits M (2010) The role of interception in the hydrological cycle. VSSD, Delft. Available at: http://resolver.tudelft.nl/uuid:7dd2523b-2169-4e7e-992c-365d2294d02e. Accessed 30 Jan 2021
24. Goodall C (2020) What we need to do now. Profile Books, London
25. Haase D (2016) Urban wetlands and riparian forests as nature-based solution for climate change adaptation in cities and their surroundings. In: Kabisch N, Korn H, Stadler J, Bonn A (eds) Nature-based solutions to climate change adaptation in urban areas. Springer Nature, Cham, pp 111–122
26. Hamway S (2020) After rocky first year, New Mexico's hemp industry poised to bloom. Available at: https://www.abqjournal.com/1413707/after-rocky-first-year-new-mexicos-hemp-industry-poised-to-bloom.html. Accessed 7 Feb 2021
27. Hillel D (1997) Small-scale irrigation for arid zones—principles and options. Food and Agriculture Organization (FAO), Rome. Available at: http://www.fao.org/3/W3094E/w3094e00.htm#TopOfPage. Accessed 29 Jan 2021
28. Hoegh-Guldberg O et al (2018) Impacts of 1.5 ºC global warming on natural and human systems. In: Masson-Delmotte V et al (eds) Global warming of 1.5 °C. An IPCC special report on the impacts of global warming of 1.5 °C above pre-industrial levels and related global greenhouse gas emission pathways. In: The context of strengthening the global response to the threat of climate change, sustainable development, and efforts to eradicate poverty. IPCC
29. Huaran H et al (2019) Fiber and seed type of hemp (*Cannabis sativa* L.) responded differently to salt-alkali stress in seedling growth and physiological indices. Ind Crops Prod 129:624–630. https://doi.org/10.1016/j.indcrop.2018.12.028
30. Huaran H, Hao L, Feihu L (2018) Seed germination of hemp (*Cannabis sativa* L.) cultivars responds differently to the stress of salt type and concentration. Ind Crops Prod 123:254–261. https://doi.org/10.1016/j.indcrop.2018.06.089

31. Iglesias AGL (2015) Adaptation strategies for agricultural water management under climate change in Europe. Agric Water Manag 155:113–124. https://doi.org/10.1016/j.agwat.2015.03.014
32. IPCC (2018) Global warming of 1.5 °C. An IPCC special report on the impacts of global warming of 1.5 °C above pre-industrial levels and related global greenhouse gas emission pathways, in the context of strengthening the global response to the threat of climate change. World Meteorological Organization, Geneva
33. Jones G, Jeliazkov VD, Roseberg RJ, Angima SD (2019) Basics of fall cover cropping for hemp in Oregon. Oregon State University, Oregon. Available at: https://catalog.extension.oregonstate.edu/sites/catalog/files/project/pdf/em9255.pdf. Accessed 28 Jan 2021
34. Kaye JP, Quemada M (2017) Using cover crops to mitigate and adapt to climate change. A review. Agron Sustain Dev 37(4):1–17. https://doi.org/10.1007/s13593-016-0410-x
35. Kebeish R et al (2007) Chloroplastic photorespiratory bypass increases photosynthesis and biomass production in *Arabidopsis thaliana*. Nat Biotech 25(5):593–599. https://doi.org/10.1038/nbt1299
36. Kraenzel DG et al (1998) Industrial hemp as an alternative crop in North Dakota. The Institute for Natural Resources and Economic Development (INRED), North Dakota. Available at: http://www.industrialhemp.net/pdf/aer402.pdf. Accessed 26 Jan 2021
37. Kurek I et al (2007) Enhanced thermostability of arabidopsis Rubisco activase improves photosynthesis and growth rates under moderate heat stress. Plant Cell 19(10):3230–3241. https://doi.org/10.1105/tpc.107.054171
38. Lalge A et al (2017) Effects of wastewater on seed germination and phytotoxicity of hemp cultivars (*Cannabis sativa* L.). In: MendelNet 2017—proceedings of 24th international PhD students conference, vol 24, pp 652–657. Available at: https://mendelnet.cz/pdfs/mnt/2017/01/125.pdf. Accessed 10 Feb 2021
39. Mateo-Sagasta J, Burke J (2011) Agriculture and water quality interactions: a global overview. SOLAW background thematic report-TR08. Food and Agriculture Organization (FAO), Rome. Available at: http://www.fao.org/3/a-bl092e.pdf. Accessed 25 Jan 2021
40. Maurer MA, Davies FS (1995) Reclaimed wastewater irrigation and rertilization of mature 'redblush' grapefruit trees on spodosols in Florida. J Am Soc Hortic Sci 120(3):394–402. https://doi.org/10.21273/JASHS.120.3.394
41. Maurino VG, Peterhansel C (2010) Photorespiration: current status and approaches for metabolic engineering. Curr Opin Plant Biol 13(3):249–256. https://doi.org/10.1016/j.pbi.2010.01.006
42. McCaskill MR et al (2016) How hail netting reduces apple fruit surface temperature: a microclimate and modelling study. Agric Meteorol 226:148–160. https://doi.org/10.1016/j.agrformet.2016.05.017
43. Mupambi G et al (2018) Protective netting improves leaf-level photosynthetic light use efficiency in 'honeycrisp' apple under heat stress. HortScience 53(10):1416–1422. https://doi.org/10.21273/HORTSCI13096-18
44. Okur B, Örçen N (2020) Soil salinization and climate change. In: Prasad MNV, Pietrzykowski M (eds) Climate change and soil interactions. Elsevier, Amsterdam, pp 331–350. https://doi.org/10.1016/b978-0-12-818032-7.00012-6
45. Oosterbaan RJ (1994) Agricultural drainage criteria. In: Ritzema HP (ed) Drainage principles and applications. ILRI Publication, Wageningen, vol 16, pp 635–688
46. Pathak H et al (1999) Soil amendment with distillery effluent for wheat and rice cultivation. Water Air Soil Pollut 113:133–140. https://doi.org/10.1023/A:1005058321924
47. Pejić B et al (2018) Effect of drip irrigation on yield and evapotranspiration of fibre hemp (*Cannabis sativa* L.). Ratar Povrt 55(3):130–134. https://doi.org/10.5937/RatPov1803130P
48. Petit J et al (2020) Genetic variability of morphological, flowering, and biomass quality traits in hemp (*Cannabis sativa* L.). Front Plant Sci 11(102):1–17. https://doi.org/10.3389/fpls.2020.00102

49. Sahara Forest Project (2019) Enabling restorative growth. Available at: https://www.saharaforestproject.com/wp-content/uploads/2019/12/Folder_liggende-A5_2019_v2_TE.pdf. Accessed 2 Feb 2021
50. Sarrantonio M, Gallandt E (2003) The role of cover crops in North American cropping systems. J Crop Prod 8(1–2):53–74. https://doi.org/10.1300/J144v08n01_04
51. Sharma L et al (2020) Plant growth-regulating molecules as thermoprotectants: functional relevance and prospects for improving heat tolerance in food crops. J Exp Bot 71(2):569–594. https://doi.org/10.1093/jxb/erz333
52. Teh SY, Koh HL (2016) Climate change and soil salinization: impact on agriculture, water and food security. Int J Agric Forest Plantation 2:1–9. Available at: https://ijafp.com/wp-content/uploads/2016/03/KLIAFP2_11.pdf. Accessed 25 Jan 2021
53. Trenberth KE et al (2007) Observations: surface and atmospheric climate change. In: Solomon S et al (eds) Climate change 2007: the physical science basis. Contribution of working group I to the fourth assessment report of the intergovernmental panel on climate change. Cambridge University Press, Cambridge, pp 235–335
54. Vessel AJ, Black CA (1947) Soil type and soil management factors in hemp production. Res Bull (Iowa Agric Home Econ Exp Stat) 28(352):381–424. Available at: http://lib.dr.iastate.edu/researchbulletin/vol28/iss352/1. Accessed 30 Jan 2021
55. Vlotman WF, Smedema LK, Rycroft DW (2020) Modern land drainage: planning, design and management of agricultural drainage systems, 2nd edn. CRC Press, Leiden
56. Vox G et al (2010) Sustainable greenhouse systems. In: Salazar A, Rios I (eds) Sustainable agriculture: technology, planning & management. Nova Science Publishers Inc., New York, pp 1–79
57. WHO (2006) WHO guidelines for the safe use of wastewater, excreta and greywater. World Health Organization (WHO), Geneva
58. Yadav RK, Kalia P, Singh SD, Varshney R (2012) Selection of genotypes of vegetables for climate change adaptation. In: Pathak H, Aggarwal P, Singh S (eds) Climate change impact, adaptation and mitigation in agriculture: methodology for assessment and application. Indian Agricultural Research Institute, New Delhi, pp 200–221
59. Zhang H et al (2018) Estimating evapotranspiration of processing tomato under plastic mulch using the SIMDualKc model. Water 10(8):1088. https://doi.org/10.3390/w10081088
60. Zinck JA, Metternicht G (2009) Soil salinity and salinization hazard. In: Zinck JA, Metternicht G (eds) Remote sensing of soil salinization: impact on land management. Taylor & Francis Group, Boca Raton, pp 3–20

Chapter 6
Future Sustainable Performance of Hemp

6.1 Climate Change Impact on Sustainable Agriculture

The meaning of sustainable development is meeting the current's necessities without compromising the ability of people in the future to fulfil their own needs [81]. The Food and Agricultural Organization [27] specifies the development process of sustainable agriculture as the utilisation of resources and the combination of environmental management with increased and continued production, secure livelihoods, food security, equity, social balance, and the cooperation of local communities. When all these conditions are met, sustainable agricultural development will be harmless to the ecosystem, technically suitable, economically viable, and socially adequate. Sustainable agriculture continues to evolve throughout the years but keeps its core multidimensional sustainability concerns economic, social, and environmental issues [70].

Siwar et al. [70] mentioned the elements of Table 6.1 as important elements for sustainable agriculture. The elements are divided between farm and policy levels. Policy level relates to regional elements that affect sustainable agriculture and extend to more than one farm.

Climate change harms sustainable agriculture. The increase in temperatures, extreme weather events, and changes in precipitation patterns results in a decreasing crop yield. As pests and crop diseases spurge under these new environmental conditions, unsustainable agricultural practices will increase. Farmers will likely react to these changes by using more fertilisers, pesticides, and irrigation water. Climate change pollution due to pesticide use degrades the sustainable performance of agriculture even more [3]. According to Agovino et al. [3], sustainable agricultural practices can only be achieved when countries work together on climate change adaptation strategies. Fine-tuning of current agricultural constraints will be less effective than the implementation of a coherent policy on a broader level. However, the action plan of these countries should not neglect current environmental constraints when seeking methods to improve agriculture productivity [58]. These policies should focus on the environmental, economic, and socio-economic level [3].

F. Dhondt and S. S. Muthu, *Hemp and Sustainability*, Sustainable Textiles: Production, Processing, Manufacturing & Chemistry,
https://doi.org/10.1007/978-981-16-3334-8_6

Table 6.1 Elements of sustainable agriculture

Farm level	Policy level
• Fertility of the soil	• Water conservation and protection
• Soil depletion	• Improve drinking water and surface water quality
• Topsoil erosion	• Protect wetlands
• Soil conservation methods	• Wildlife habitat and increased amounts of beneficial insects
• Reduced tillage and zero tillage	• Diversity of wildlife
• Groundwater pollution	• Decrease of family farms
• Prevent erosion of soil due to wind and water flows	• The collapse of economic and social circumstances in rural communities
• Intercropping	• Contribution of the farms to local communities
• Growing a variety of crops to reduce risks from extreme weather events	• Product value
• Pest management	• Social/human capital
• Nitrogen fertilisation and reduction of purchased fertiliser costs	• Local economy
• Agro-forestry	• Increasing costs of production
• Living and working conditions for farm employees	• Food and agricultural in social and political policies
• Long-term yield	• National ethics
• Long-term profitability	• Rural community development
• Land use	• Market conditions
• Labour	• Pace of environmental degradation
• Innovative selling through marketing strategies	• Running community-supported agriculture farms
• Mindful use of renewable and recyclable resources	

1. Environmental policies should be more consistent to encourage sustainable agriculture. Some important focus points, among others, are investments in research and development, correct insurance subsidies, large strategic portfolios, including adaptation and mitigation strategies [57].
2. The economic level of agriculture is currently held back by environmental constraints. Pest controlling techniques and emission substitutes allow farmers to increase their weight trade-offs and productivity growth instead of restricting them [3].
3. On a socio-economic level, innovation, macroeconomic trade, investment, infrastructure, education, training, and rural incomes are essential [3].

Not every adaptation method is accessible for every farmer as there are large economic differences in the countries where hemp is grown. Hemp crops are cultivated in approximately 47 countries for commercial or research purposes [66].

Currently, fertilisation use and fertilisation production contributed for a large part to the environmental impact of hemp crops. The main nutrients that are used in hemp production are nitrogen and phosphorus [30, 75]. Another impacting factor in hemp crop production is diesel production and use for agricultural machinery [75].

This chapter will assess the possible climate change adaptation methods and will go more in-depth on the likelihood of pests and disease occurrence for hemp crops. Besides the negative climate change impacts, some crop processes benefit from changing environmental conditions.

6.2 Photosynthesis

Climate change has an impact on crop photosynthesis through increased greenhouse gasses in the atmosphere. An increase in greenhouse gasses in the atmosphere increases the number of aerosols and carbon dioxide uptake of crops. Photosynthesis is a process where the energy of light in combination with carbon dioxide transforms into sugars. As aerosols change the amount of solar radiation that reaches the surfaces, it influences the photosynthesis process. Solar radiation will reach the earth more scattered if there are many aerosols in the atmosphere. Diffused radiation will likely increase the photosynthetic activities of crops [35, 53, 83]. The increased amount of carbon dioxide in the atmosphere also influences the photosynthesis process of C3 crops. The photosynthetic rates increased together with crop growth [20, 25, 39]. As hemp is a C3 crop, it will likely have increased photosynthetic rates as well, with an increase of CO_2 [15]. The productive photosynthetic rates are not only dependent on the carbon dioxide rates, but other favourable climate conditions, such as optimal temperature, water, and nutrient concentrations, should be present as well [25, 40].

Increased photosynthesis in crops increases the carbon dioxide uptake and increases crop growth with the same input of water and nutrients. However, the crop should grow in optimal climate conditions to benefit from increased carbon dioxide [25, 40]. A crop grown in optimal climate conditions with increased carbon dioxide rates and diffused light results in higher crop growth for the same number of inputs and can be more sustainable. However, it should be noted that an increased amount of greenhouse gasses causes many other environmental impacts and health issues and should therefore not be seen as a possible sustainable growth booster for crops.

6.3 Temperature

Climate change will likely increase global temperatures [41]. As every crop has an optimum, minimum and maximum temperature, a change in temperature will impact crop yield. Rising temperatures reduce the productivity of hemp crop growth

[80] and can cause an increase in pest and diseases [37]. Besides the impact on the crop, the fibre yield and quality will decrease when temperatures exceed the minimum or maximum temperature of hemp crops [11, 18, 32].

Possible adaptation methods to sustain hemp growth are a change in genotype, the use of hail netting or the use of plant growth regulators. Genetic engineering is a method in which the optimum heat tolerance traits or other preferred genes of hemp can be selected for crop breeding [84]. Hail netting reduces the amount of direct sunlight and lowers soil temperatures [49, 56]. Plant growth regulators increase the heat tolerance of hemp through organic substances that control the development of hemp and create a defence system [68]. The adaptation methods could adapt the crop to changing temperatures and with that avoid reduced fibre quality, fibre yield, and crop mortality. However, these adaptation methods have their own environmental impact as well.

Genetic engineering can improve crops tolerance against environmental stresses and excessive heat [7]. Besides all the positive impacts of genetic engineering, there are also negative consequences of the use of genetically engineered crops. Some risks of genetic engineering are genetic contamination, competition with natural species, transfer of genes to other organisms, unpredictable or unintended effects, and ecosystem impacts [61]. Genetic engineering might harm the biodiversity in certain regions as only the same crop would be grown in a region [4]. Hemp scores high on biodiversity [54], but genetic engineering can affect this by decreasing biodiversity in the crop's regions. Biodiversity benefits from agriculture that makes use of different species and crop diversity. The diversity of crops results in pollinator habitats and protection from diseases and pests, and increase ecosystem qualities. To avoid negative consequences of genetically engineered crops, farmers should incorporate crop diversification and rotations [4]. Genetic engineering can also contribute to biodiversity when engineered crops intersect with wild species and increase the wild species resistance [10]. Nevertheless, Prakash et al. [61] marked the intersection between wild species and engineered crops as a risk, as certain crop traits might disappear due to these intersections.

Another adaptation method, hail netting, is often made of plastic petroleum-based materials and thus not the most sustainable option. Plastic nets are a leading source of agricultural waste [79]. Furthermore, excessive solar radiation and heat in combination with plastic nets can release toxic substances [42]. The use of bio-based plastics can reduce the environmental impact of hail netting. Bioplastic is made of renewable agricultural biomass sources, and most biopolymers are biodegradable. The biodegradability rate of bioplastics can differ according to the stereochemistry, length of side chains, and the sensitivity to hydrolysis. The need for durable biopolymers might outdo the biodegradability of the material; therefore, the end-of-life of hail nets needs to be thought out [55].

Plant growth regulators are organic substances that control the growth of crops and protect the crop against environmental stresses [36, 68]. Plant growth regulators also reduce the need for pesticides as the defence system, and tolerance of crops improves [36]. An increase in temperature can increase the need for pesticides as pest and diseases increase [37]. Currently, hemp does not require pesticides, so this would lead to a higher environmental impact. Besides temperature stress, plant growth regulators are also a crucial adaptation method for crops grown under moisture stress [38].

High temperatures can also increase the THC level of hemp crops [69]. When THC levels exceed the local law's norm, the hemp yields will be destroyed. This would be a waste of inputs and energy use from machinery.

6.4 Water Availability

Climate change influences water availability due to changing temperatures, rain patterns, wind speed, vegetation cover, soil moisture, and runoff [72]. The availability of water is bound to the geographical location and its other associated factors. In some region's precipitation and torrential rain increases, whereas in other regions, there will be a deficiency of water [41]. The availability of water is vital for hemp crops in the first six weeks, whereas more mature crops can better withstand dry conditions [74]. Water stress decreases hemp yield and fibre quality [2, 6, 16, 24]. Excessive rain combined with clogged soil reduces hemp yield and increases hemp fungal diseases [14, 33, 51]. Hail can also damage hemp crops [60].

There are multiple adaptation measures for both droughts and floods. Possible adaptation methods to combat water stress are salt distillation, wastewater use, water use efficiency, genetic engineering, improved irrigation, change in cropping patterns. Possible methods to avoid an excessive amount of water on the land and to combat floods are rainfall interception, wetlands, and agricultural drainage systems. Sustainable adaptation methods that will not have a negative impact on environmental or social sustainability are the use of rainfall interception, organic mulching, and growing crops that are well adapted to the local climate conditions.

The innovation of the Sahara Forest Project uses saltwater and extracts the salt from the water, which results in freshwater. This water can be used to irrigate crops in water-stressed areas. Even though this innovation would be a good outcome for a future under rapidly changing climate conditions, the energy use and costs of this innovation a relatively high. The Sahara Forest Project uses solar panels instead of fossil fuel energy sources [63]. The use of solar panels is a more sustainable solution. However, desalination of seawater, as well as solar panels, is too expensive for farmers in developing countries [67]. Water stress is and will be a rather large problem in developing areas, so it is important that future innovations take into account the needs of farmers in those countries [71].

Treated wastewater is another innovative way to use water for irrigation. Farmers treat the wastewater before using it on agricultural land [82]. Hemp crops benefit from the nutrients in the wastewater [43]. However, wastewater treatments are not accessible for every farmer [29, 62]. Especially in developing countries, untreated wastewater is used to irrigate farmlands. Untreated wastewater contains chemicals and other substances that can harm the health of local populations and the environment [62]. The global warming potential of wastewater treatments in developing countries is higher than in developed countries, whereas the electricity consumption is lower in developing countries. More detailed life cycles assessments are necessary to measure the environmental impact of wastewater treatments in developing countries. The social impact of wastewater use and treatment is positive, as it increases the amount of clean water and sanitation for local communities [29].

As water can be scarce in certain areas, more farmers will need to irrigate their land [50]. Especially in water-stressed areas, this can result in an even larger water deficit for local communities [65]. Even though hemp requires less water than cotton per ton of fibre [52], irrigation can still impact the groundwater levels when land is not irrigated properly [65]. Both drip irrigation and deficit irrigation can be an outcome to prevent the depletion of water [5, 13, 28, 87]. These more efficient irrigation methods are slightly more susceptible to climate change than other irrigation methods. The impacts of climate change on irrigation efficiency further need to be studied on the water use efficiency of future irrigation methods [46]. Using groundwater instead of rainwater will increase the blue water footprint of hemp, whereas the green water footprint will reduce. Blue water is the amount of groundwater that is consumed, and green water refers to the amount of consumed rainwater [52].

Farmers can use wetlands as a flood adaptation method. Wetlands increase biodiversity and the resistance of the land against high temperatures [31]. This method is used in between the crop growth periods, as hemp cannot be grown for a longer period in wet soils [24, 78]. Where on the one hand, wetlands store large amounts of carbon dioxide. On the other hand, wetlands emit a high amount of greenhouse gas emissions, CH_4 and N_2O [23]. Another mentioned impact of wetlands is that it is a breeding place for disease-carrying parasites [8].

Hemp is more prone to diseases in long rain periods and when floods on agricultural land occur. The fungal diseases *Botrytis cinerea* and *Sclerotinia sclerotiorum* [51] can be overcome by either changing to a genotype that is less prone to these diseases [76] or by the use of fungicides [22]. As crop diseases will likely increase under changing climate conditions [48], there is a great chance that hemp farmers will use more fungicides to protect their harvest. The use of fungicides will increase the environmental footprint of cultivating hemp. Chemical fungicides pollute water and are toxic to a large range of organisms [88]. Climate change will furthermore increase the number of chemicals that reach local communities through increased use [12].

Improving moisture in the soil and preventing flooding can be accomplished by proper soil health. The next subchapter will go more in-depth on the sustainability of those adaptation methods and the possible consequences of climate change on the sustainability of soil health.

6.5 Soil

6.5.1 Cover Cropping and Crop Rotation

Cover cropping and crop rotation improve the environmental impact of crops as it increases organic matter, soil moisture, water infiltration and decreases nutrient loss and soil erosion [21, 64]. Cover crops furthermore reduce nitrogen leaching and, with that, reduces environmental nitrogen pollution [64]. Soil diseases are reduced as microbial life improves, which reduces the number of needed herbicides and pesticides for hemp production [17]. Overall, these two methods improve the performance and impact of hemp.

6.5.2 Mulch Cropping

Mulch cropping is an adaptation method that lowers soil temperature, erosion, improves soil fertility, soil structure and reduces weed growth. Mulch covers the soil and can be either crop residues or plastic mulch [84]. Crop residues are a sustainable adaptation method that will not further impact hemp's performance. Plastic mulch, on the other hand, further increases agricultural plastic pollution and decreases long-term crop yield [86]. Therefore, farmers should choose organic crop residues over inorganic plastic mulch. Mulching is an adaptation method that is accessible to many farmers as it is affordable [84].

6.5.3 Soil Water Leaching

Climate change will intensify salinisation in soils. Salinisation leads to reduced hemp stem and root lengths [34]. Soil water leaching can lower the salt content in soils and includes flushing, irrigation and drainage [59]. Soil water leaching uses a large amount of water. Soil water leaching increases the amount of leaching soil chemicals which increases the risk of groundwater contamination [45]. The increased use of water for soil leaching increases the water footprint of hemp crops. Besides that, soil water flushing is not an available option for most farmers as it is too pricey [19].

6.5.4 Fertilisation Management

As temperatures increase, the agricultural fertilisation inputs increase with it. Farmers try to mitigate crop failure by changes in production inputs that increase agricultural emissions [26]. The environmental impact of hemp production is for a large part due to fertilisation use [1, 30, 73, 75, 77, 85]. Increased inputs due to climate change will increase the environmental impact of hemp if fertilisation is not managed properly. Farmers should adequately manage fertilisation inputs to reduce agricultural emissions [26]. Another way to decrease fertilisation emissions and pollution is to change to bio-fertilisers [9]. Furthermore, decreased fertilisation inputs reduce the possibility of salinisation in soils, as fertilisers contain salts as well [47]. Excess nitrogen use also leads to hemp stems that stay green for a longer period of time which leads to longer drying times and complications in harvesting and fibre processing [44].

6.6 Conclusion

Climate change will challenge the availability of raw materials and will impact temperatures, water resources, and soil health. As a response to these environmental stresses, the agricultural sector came up with adaptation methods. However, with each adaptation method that is implemented, the sustainability of that specific adaptation should be considered as well. Some climatic changes require a change in pesticides, fertilisers, and fungicides. The use of these chemicals, if they are inorganic, can further damage the environment and further contribute to climate change. Furthermore, adaptation methods are often only focused on the environmental perspective but often overlook the impact on local communities. Many of the mentioned adaptation methods in this chapter are not accessible for farmers in developing countries as those methods are too expensive.

References

1. Abass E (2005) Life cycle assessment of novel hemp fibre—a review of the green decortication process. Imperial College London, Department of Environmental Science and Technology, London
2. Abot A et al (2013) Effects of cultural conditions on the hemp (*Cannabis sativa*) phloem fibres: biological development and mechanical properties. J Compos Mater 8(47):1067–1077. https://doi.org/10.1177/0021998313477669
3. Agovino M et al (2019) Agriculture, climate change and sustainability: the case of EU-28. Ecol Ind 105:525–543. https://doi.org/10.1016/j.ecolind.2018.04.064
4. Aguilar J et al (2015) Crop species diversity changes in the United States: 1978–2012. PLoS ONE 10(8):e0136580. https://doi.org/10.1371/journal.pone.0136580
5. Ali MH, Hoque MR, Hassan AA, Khair A (2007) Effects of deficit irrigation on yield, water productivity, and economic returns of wheat. Agric Water Manag 92(3):151–161. https://doi.org/10.1016/j.agwat.2007.05.010
6. Amaducci S, Zatta A, Pelatti F, Venturi G (2008) Influence of agronomic factors on yield and quality of hemp (*Cannabis sativa* L.) fibre and implication for an innovative production system. Field Crops Res 107(2):161–169. https://doi.org/10.1016/j.fcr.2008.02.002

7. Anderson R, Bayer PE, Edwards D (2020) Climate change and the need for agricultural adaptation. Curr Opin Plant Biol 56:197–202. https://doi.org/10.1016/j.pbi.2019.12.006
8. Andersson E, Borgström S, McPhearson T (2016) Double insurance in dealing with extremes: ecological and social factors for making nature-based solutions to last. In: Kabisch N, Korn H, Stadler J, Bonn A (eds) Nature-based solutions to climate chagne adaptation in urban areas. Springer Nature, Cham, pp 51–64
9. Arora NK (2019) Impact of climate change on agriculture production and its sustainable solutions. Environ Sustain 2:95–96. https://doi.org/10.1007/s42398-019-00078-w
10. Barrows G, Sexton S, Zilberman D (2014) Agricultural biotechnology: the promise and prospects of genetically modified crops. J Econ Perspect 28(1):99–120. https://doi.org/10.1257/jep.28.1.99
11. BCMAF (1999) Industrial hemp (*Cannabis sativa* L.) factsheet. British Colombia Ministry of Agriculture and Food, Kamploops. Available at: https://www.votehemp.com/wp-content/uploads/2018/09/hempinfo.pdf. Accessed 10 Dec 2020
12. Boxall AB et al (2009) Impacts of climate change on indirect human exposure to pathogens and chemicals from agriculture. Environ Health Perspect 117(4):508–514. https://doi.org/10.1289/ehp.0800084
13. Brouwer C, Prins K, Kay M, Heilbloem M (1988) Irrigation water management: irrigation methods. Food and Agriculture Organization (FAO), Rome. Available at: http://www.fao.org/3/s8684e/s8684e00.htm#Contents. Accessed 29 Jan 2021
14. Canadian Hemp Trade Alliance (2020) Impacts of severe weather events on hemp production. Available at: http://www.hemptrade.ca/eguide/production/impacts-of-severe-weather-events-on-hemp-production. Accessed 4 Dec 2020
15. Chandra S, Lata H, Khan IA, Elsohly MA (2008) Photosynthetic response of *Cannabis sativa* L. to variations in photosynthetic photon flux densities, temperature and CO2 conditions. Physiol Mol Biol Plants 14(4):299–306. https://doi.org/10.1007/s12298-008-0027-x
16. Chemikosova SB, Pavlencheva NV, Gur'yanov OP, Gorshkova TA (2006) The effect of soil drought on the phloem fiber development in long-fiber flax. Russ J Plant Physiol 53(5):656–662. https://doi.org/10.1134/S1021443706050098
17. Clark A (2008) Managing cover crops profitably, 2nd edn. Sustainable Agriculture Network, Beltsville
18. Craufurd PQ, Wheeler TR (2009) Climate change and the flowering time of annual crops. J Exp Bot 60(9):2529–2539. https://doi.org/10.1093/jxb/erp196
19. Cuevas J et al (2019) A review of soil-improving cropping systems for soil salinization. Agronomy 9(295):1–22. https://doi.org/10.3390/agronomy9060295
20. Cure JD, Acock B (1986) Crop responses to carbon dioxide doubling: a literature survey. Agric For Meteorol 38(1–3):127–145. https://doi.org/10.1016/0168-1923(86)90054-7
21. de Azevedo DMP, Landivar J, Vieira RM, Moseley D (1999) The effect of cover crop and crop rotation on soil water storage and on sorghum yield. Pesq Agrop Brasileira 34(3):391–398. https://doi.org/10.1590/s0100-204x1999000300010
22. de Figueirêdo GS et al (2010) Biological and chemical control of Sclerotinia sclerotiorum using Trichoderma spp. and Ulocladium atrum and pathogenicity to bean plants. Braz Arch Biol Technol 53(1):1–9. https://doi.org/10.1590/S1516-89132010000100001
23. de Klein JJM, van der Werf AK (2014) Balancing carbon sequestration and GHG emissions in a constructed wetland. Ecol Eng 66:36–42. https://doi.org/10.1016/j.ecoleng.2013.04.060
24. Dewey LH (1914) The yearbook of the United States department of agriculture 1913. U.S. Department of Agriculture, Washington, D.C.
25. Drake BG, Gonzalez-Meler MA, Long SP (1997) More efficient plants: a consequence of rising atmospheric CO2? Annu Rev Plant Physiol Plant Mol Biol 48:609–639. https://doi.org/10.1146/annurev.arplant.48.1.609
26. Erbas BC, Solakoglu EG (2017) In the presence of climate change, the use of fertilizers and the effect of income on agricultural emissions. Sustainability 9:1989. https://doi.org/10.3390/su9111989

27. FAO (1989) The state of food and agriculture. Food and Agricultural organization (FAO), Rome. Available at: http://www.fao.org/3/t0162e/t0162e.pdf. Accessed 1 Mar 2021
28. FAO (2013) Climate-smart agriculture sourcebook. Food and Agriculture Organization (FAO), Rome. Available at: http://www.fao.org/3/i3325e/i3325e.pdf. Accessed 3 Feb 2021
29. Gallego-Schmid A, Zepon Tarpani RR (2019) Life cycle assessment of wastewater treatment in developing countries: a review. Water Res 153:63–79. https://doi.org/10.1016/j.watres.2019.01.010
30. González-García S, Hospido A, Feijoo G, Moreira MT (2010) Life cycle assessment of raw materials for non-wood pulp mills: hemp and flax. Resour Conserv Recycl 54(11):923–930. https://doi.org/10.1016/j.resconrec.2010.01.011
31. Haase D (2016) Urban wetlands and riparian forests as nature-based solution for climate change adaptation in cities and their surroundings. In: Kabisch N, Korn H, Stadler J, Bonn A (eds) Nature-based solutions to climate change adaptation in urban areas. Springer Nature, Cham, pp 111–122
32. Hall J, Bhattarai SP, Midmore DJ (2012) Review of flowering control in industrial hemp. J Nat Fibers 9(1):23–36. https://doi.org/10.1080/15440478.2012.651848
33. Harper JK et al (2018) Industrial hemp production. The Pennsylvania State University, Pennsylvania. Available at: https://extension.psu.edu/industrial-hemp-production. Accessed 6 Dec 2020
34. Huaran H et al (2019) Fiber and seed type of hemp (*Cannabis sativa* L.) responded differently to salt-alkali stress in seedling growth and physiological indices. Ind Crops Prod 129:624–630. https://doi.org/10.1016/j.indcrop.2018.12.028
35. IPCC (2013) Climate change 2013: the physical science basis. Contribution of working group I to the fifth assessment report of the intergovernmental panel on climate change. Cambridge University Press, Cambridge
36. Jan S et al (2020) Plant growth regulators: a sustainable approach to combat pesticide toxicity. 3 Biotech 10(466). https://doi.org/10.1007/s13205-020-02454-4
37. Karmakar R, Das I, Dutta D, Rakshit A (2016) Potential effects of climate change on soil properties: a review. Sci Int 4(2):51–73. https://doi.org/10.17311/sciintl.2016.51.73
38. Khan N, Bano AMD, Babar A (2020) Impacts of plant growth promoters and plant growth regulators on rainfed agriculture. PLoS ONE 15(4):e0231426. https://doi.org/10.1371/journal.pone.0231426
39. Kimball BA (1983) Carbon dioxide and agricultural yield: an assemblage and analysis of 430 prior observations. Agron J 75(5):779–788. https://doi.org/10.2134/agronj1983.00021962007500050014x
40. Kirschbaum MUF (2004) Direct and indirect climate change effects on photosynthesis and transpiration. Plant Biol 6:242–253. https://doi.org/10.1055/s-2004-820883
41. KNMI and WMO (2020) KNMI climate change atlas. Available at: https://climexp.knmi.nl/plot_atlas_form.py. Accessed 24 Jan 2021
42. Lackner M (2015) Bioplastics—biobased plastics as renewable and/or biodegradable alternatives to petroplastics. In: Othmer K (ed) Encyclopedia of chemical technology. Wiley, pp 1–41
43. Lalge A et al (2017) Effects of wastewater on seed germination and phytotoxicity of hemp cultivars (*Cannabis sativa* L.). In: MendelNet 2017—proceedings of 24th international PhD students conference, vol 24, pp 652–657. Available at: https://mendelnet.cz/pdfs/mnt/2017/01/125.pdf. Accessed 10 Feb 2021
44. Legros S, Picault S, Cerruti N (2013) Factors affecting the yield of industrial hemp—experimental results from France. In: Bouloc P, Allegret S, Arnaud L (eds) Hemp: industrial production and uses. CAB International, Wallingford, pp 72–97
45. Letey J et al (2011) Evaluation of soil salinity leaching requirement guidelines. Agric Water Manag 98(4):502–506. https://doi.org/10.1016/j.agwat.2010.08.009
46. Malek K, Adam JC, Stöckle CO, Peters RT (2018) Climate change reduces water availability for agriculture by decreasing non-evaporative irrigation losses. J Hydrol 561:444–460. https://doi.org/10.1016/j.jhydrol.2017.11.046

47. Mateo-Sagasta J, Burke J (2011) Agriculture and water quality interactions: a global overview. SOLAW background thematic report-TR08. Food and Agriculture Organization (FAO), Rome. Available at: http://www.fao.org/3/a-bl092e.pdf. Accessed 25 Jan 2021
48. Mbow C et al (2019) Food security. In: Shukla PR et al (eds) Climate change and land: an IPCC special report on climate change, desertification, land degradation, sustainable land management, food security, and greenhouse gas fluxes in terrestrial ecosystem. IPCC, Geneva, pp 437–550
49. McCaskill MR et al (2016) How hail netting reduces apple fruit surface temperature: a microclimate and modelling study. Agric For Meteorol 226–227:148–160. https://doi.org/10.1016/j.agrformet.2016.05.017
50. McDonald RI, Girvetz EH (2013). Two challenges for U.S. irrigation due to climate change: increasing irrigated area in wet states and increasing irrigation rates in dry states. PLoS ONE 8(6): e65589. https://doi.org/10.1371/journal.pone.0065589
51. Meijer WJM, van der Werf HMG, Mathijssen EWJM, van den Brink PWM (1995) Constraints to dry matter production in fibre hemp (*Cannabis sativa* L.). Eur J Agron 4(1): 109–117. https://doi.org/10.1016/S1161-0301(14)80022-1
52. Mekonnen MM, Hoekstra AY (2011) The green, blue and grey water footprint of crops and derived crop products. Hydrol Earth Syst Sci 15:1577–1600. https://doi.org/10.5194/hess-15-1577-2011
53. Mercado LM et al (2009) Impact of changes in diffuse radiation on the global land carbon sink. Nature 458:1014–1018. https://doi.org/10.1038/nature07949
54. Montford S, Small E (1999) A comparison of the biodiversity friendliness of crops with special reference to hemp (*Cannabis sativa* L.). J Int Hemp Assoc 6(2):53–63. Available at: http://www.internationalhempassociation.org/jiha/jiha6206.html. Accessed 31 Oct 2020
55. Mukherjee A, Knoch S, Tavares J (2019) Use of bio-based polymers in agricultural exclusion nets: a perspective. Biosyst Eng 180:121–145. https://doi.org/10.1016/j.biosystemseng.2019.01.017
56. Mupambi G et al (2018) Protective netting improves leaf-level photosynthetic light use efficiency in 'honeycrisp' apple under heat stress. HortScience 53(10):1416–1422. https://doi.org/10.21273/HORTSCI13096-18
57. OECD (2012) Farmer behaviour, agricultural management and climate change. OECD, Paris. https://doi.org/10.1787/9789264167650-en
58. OECD (2016) Agriculture and climate change: towards sustainable, productive and climate-friendly agricultural systems. OECD, Paris. Availabe at: https://www-oecd-org.proxy.library.uu.nl/agriculture/ministerial/background/notes/4_background_note.pdf. Accessed 2 Mar 2021
59. Okur B, Örçen N (2020) Soil salinization and climate change. In: Prasad MNV, Pietrzykowski M (eds) Climate change and soil interactions. Elsevier, Amsterdam, pp 331–350. https://doi.org/10.1016/b978-0-12-818032-7.00012-6
60. Pahkala K, Pahkala E, Syrjälä H (2008) Northern limits to fiber hemp production in Europe. J Ind Hemp 13(2):104–116. https://doi.org/10.1080/15377880802391084
61. Prakash D, Verma S, Bhatia R, Tiwary BN (2011) Risks and precautions of genetically modified organisms. ISRN Ecol 2011:1–13. https://doi.org/10.5402/2011/369573
62. Qadir M et al (2010) The challenges of wastewater irrigation in developing countries. Agric Water Manag 97(4):561–568. https://doi.org/10.1016/j.agwat.2008.11.004
63. Sahara Forest Project (2019) Enabling restorative growth. Available at: https://www.saharaforestproject.com/wp-content/uploads/2019/12/Folder_liggende-A5_2019_v2_TE.pdf. Accessed 2 Feb 2021
64. Sarrantonio M, Gallandt E (2003) The role of cover crops in North American cropping systems. J Crop Prod 8(1–2):53–74. https://doi.org/10.1300/J144v08n01_04
65. Scanlon BR et al (2012) Groundwater depletion and sustainability of irrigation in the US High Plains and Central Valley. PNAS 109(24):9320–9325. https://doi.org/10.1073/pnas.1200311109

66. Schluttenhofer C, Yuan L (2017) Challenges towards revitalizing hemp: a multifaceted crop. Trends Plant Sci 22(11):917–929. https://doi.org/10.1016/j.tplants.2017.08.004
67. Shahsavari A, Akbari M (2018) Potential of solar energy in developing countries for reducing energy-related emissions. Renew Sustain Energy Rev 90:275–291. https://doi.org/10.1016/j.rser.2018.03.065
68. Sharma L et al (2020) Plant growth-regulating molecules as thermoprotectants: functional relevance and prospects for improving heat tolerance in food crops. J Exp Bot 71(2):569–594. https://doi.org/10.1093/jxb/erz333
69. Sikora V, Berenji J, Latković D (2011) Influence of agroclimatic conditions on content of main cannabinoids in industrial hemp (*Cannabis sativa* L.). Genetika 43(3):449–456. https://doi.org/10.2298/GENSR1103449S
70. Siwar C, Alam MM, Murad MW, Al-Amin AQ (2009) A review of the linkages between climate change, agricultural sustainability and poverty in Malaysia. Int Rev Bus Res Pap 5 (6):309–32. Available at: https://papers.ssrn.com/sol3/papers.cfm?abstract_id=2941285. Accessed 1 Mar 2021
71. Tan X et al (2021) Research on the status and priority needs of developing countries to address climate change. J Cleaner Prod 289:125669. https://doi.org/10.1016/j.jclepro.2020.125669
72. Trenberth KE et al (2007) Observations: surface and atmospheric climate change. In: Solomon S et al (eds) Climate change 2007: the physical science basis. Contribution of working group I to the fourth assessment report of the intergovernmental panel on climate change. Cambridge University Press, Cambridge, pp 235–335
73. Turunen L, van der Werf HMG (2006) Life cycle analysis of hemp textile yarn. Comparison of three fibre processing scenarios and a flax scenario. INFRA, UMR SAS, Rennes
74. USDA (1914) The yearbook of the United States department of agriculture 1913. U.S. Department of Agriculture, Washington, D.C.
75. van der Werf HMG (2004) Life cycle analysis of field production of fibre hemp, the effect of production practices on environmental impacts. Euphytica 140(1):13–23. https://doi.org/10.1007/s10681-004-4750-2
76. van der Werf HMG, van Geel WCA, Wijlhuizen M (1995) Agronomic research on hemp (*Cannabis sativa* L.) in the Netherlands, 1987–1993. J Int Hemp Assoc 2(1):14–17. Available at: http://druglibrary.net/olsen/HEMP/IHA/iha02107.html. Accessed 15 Dec 2020
77. van Eynde H (2015) Comparative life cycle assessment of hemp and cotton fibres used in Chinese textile manufacturing. KU Leuven, Leuven
78. Vessel AJ, Black CA (1947) Soil type and soil management factors in hemp production. Res Bull (Iowa Agric Home Econ Exp Stat) 28(352):381–424 (Article 1). Available at: http://lib.dr.iastate.edu/researchbulletin/vol28/iss352/1. Accessed 30 Jan 2021
79. Vox G et al (2016) Mapping of agriculture plastic waste. Agric Agric Sci Procedia 8:583–591. https://doi.org/10.1016/j.aaspro.2016.02.080
80. Wallace-Wells D (2019) The uninhabitable earth a story of the future, 1st edn. Penguin Books, London
81. WCED (1987) Our common future. Oxford University Press, Oxford
82. WHO (2006) WHO guidelines for the safe use of wastewater, excreta and greywater. World Health Organization (WHO), Geneva. Available at: https://www.who.int/water_sanitation_health/publications/gsuweg4/en/. Accessed 3 Feb 2021
83. Wild M (2009) Global dimming and brightening: a review. J Geophys Res 114. https://doi.org/10.1029/2008JD011470
84. Yadav RK, Kalia P, Singh SD, Varshney R (2012) Selection of genotypes of vegetables for climate change adaptation. In: Pathak H, Aggarwal P, Singh S (eds) Climate change impact, adaptation and mitigation in agriculture: methodology for assessment and application. Indian Agricultural Research Institute, New Delhi, pp 200–221
85. Zampori L, Dotelli G, Vernelli V (2013) Life cycle assessment of hemp cultivation and use of hemp-based thermal insulator materials in buildings. Environ Sci Technol 47(13):7413–7420. https://doi.org/10.1021/es401326a

86. Zhang D et al (2020) Plastic pollution in croplands threatens long-term food security. Glob Change Biol 26:3356–3367. https://doi.org/10.1111/gcb.15043
87. Zhang H et al (2018) Estimating evapotranspiration of processing tomato under plastic mulch using the SIMDualKc model. Water 10(8):1088. https://doi.org/10.3390/w10081088
88. Zubrod JP et al (2019) Fungicides: an overlooked pesticide class? Environ Sci Technol 53:3347–3365. https://doi.org/10.1021/acs.est.8b04392

Chapter 7
Sustainable Hemp Products

7.1 Hemp Industries

Interest in hemp products is growing as the need for more products with a lower environmental footprint increases. Hemp as a low-input, fast-growing, and high-yield crop [30, 43, 75] has a great potential for future sustainable products. Hemp can be used in a broad range of industries [16].

Every part of hemp can be used for different end-uses. Most of the times, hemp is either grown for its fibres, seeds, cannabidiol (CBD) or as a dual-purpose crop. Hemp seeds can be used in food products, cosmetics, and animal feed [28, 51, 55, 74]. The fibres are most often processed into textiles, paper, composite, and insulation [16, 39, 66]. Leaves, roots, and hurds of the crop are by-products of hemp yield, which can be used again in other products. The leaves can be used for medicine, tea, and nutrient purposes [17, 53]. The hurds is the wooden part of hemp stalks, which are not suitable for textile purposes but can be processed into building materials, insulation, garden mulch, paper, composite, animal bedding, and acoustic panels [24, 44, 47, 86, 67]. The roots can be processed into pulp that is suitable for paper products [62]. The only industry in which hemp pulp currently is used commercially is the cigarette paper industry [16]. An overview of hemp uses is displayed in Fig. 7.1.

7.2 Current Industries

The current industries in which hemp is used are the construction and insulation sector, food and nutrition, paper processing, and the textile industry [16]. Hemp is an interesting material for these industries as these sectors are seeking sustainable products that reduce the environmental impacts of the current operations. In the construction and insulation sector, hemp is often mixed with lime, clay, or cement

F. Dhondt and S. S. Muthu, *Hemp and Sustainability*, Sustainable Textiles: Production, Processing, Manufacturing & Chemistry,
https://doi.org/10.1007/978-981-16-3334-8_7

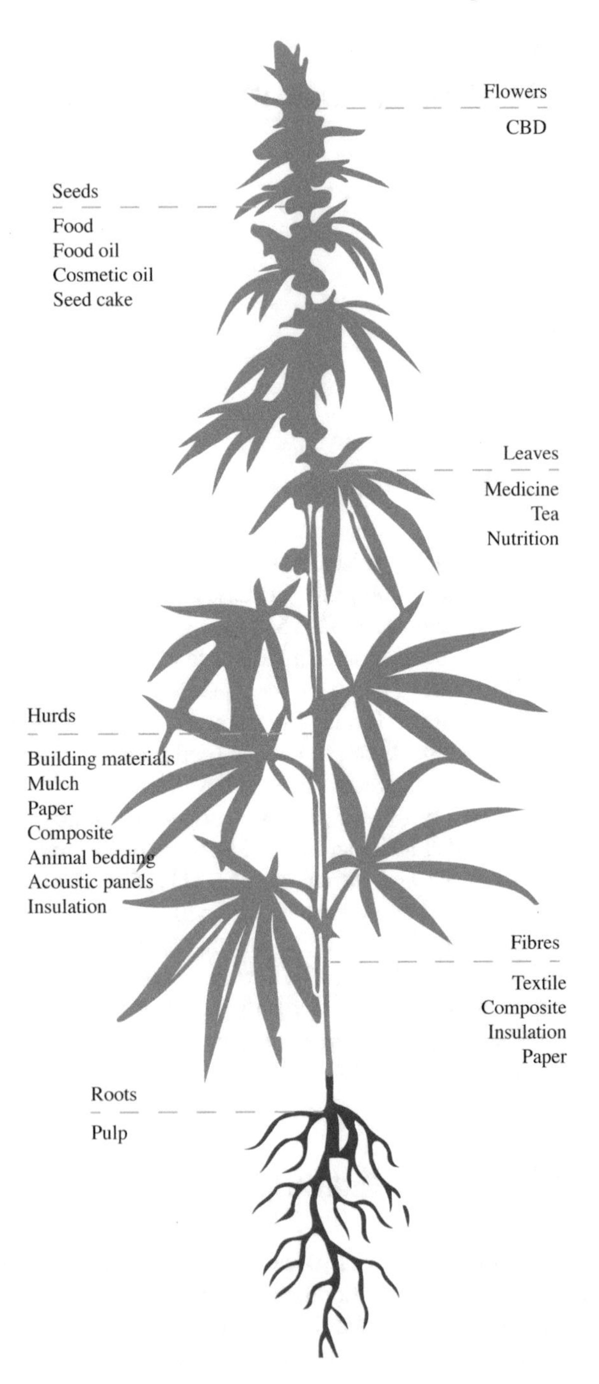

Fig. 7.1 Different uses of the hemp crop

to form building blocks. But the crop can be used on itself, without being mixed with other materials, for insulation, food, paper, and textile. The seeds of the hemp crop can be processed into food products, food oils, and animal feed. Hemp and textile use dates far back as hemp was already cultivated in China more than 6000 years ago [94]. Paper remaining's have been found that date back to 2685 to 2138 a BC and textile remaining's that were from seventh-century BC [2, 6].

7.2.1 Construction and Insulation

Eco-friendly building materials such as hemp concrete, hemp concrete blocks, and hemp shive insulation are becoming more popular for construction projects [15]. Building materials are defined as eco-friendly when the environmental impact and heating and cooling energy consumption for the building are low, and the indoor environmental quality with regard to safety and comfort is high [14]. The different types of building materials that can be made out of hemp are extensive. The bast fibres can be used for insulation [16]. The other part of the stalks, the wooden core, is used for hurds that can be mixed with lime, clay, or cement to form building blocks [24]. The building materials that can be made from hemp materials are bricks, slabs, panels, wallboards, fibreboards, and roofing tiles [16].

The main benefits of using hemp instead of concrete, wood, or plastics are the reduced energy required for production and the sequestration of carbon by hemp building materials [76]. Other interesting properties of hemp building materials are that the materials are lightweight, have low thermal conductivity [25, 49], are exceptional hydric regulators [14], have low production costs [16], are waterproof and moisture-proof [59]. Some downsides of using hemp fibres as a building material are that the durability cannot be guaranteed, and the material is sensitive to mould in humid environments [59]. More research into the strength, durability, and fire resistance of hemp concrete is needed before it can be fully implemented in the fast-growing green building industry [42].

7.2.2 Food and Nutrition

As the global protein demand and interest for healthy foods and dietary supplements will increase in the future, the food industry searches for alternative food crops [16, 46]. Hemp seeds are a rich source of fat and proteins for animal and human food [74]. Hemp food is only made from hemp seeds. The seeds contain 30% oil and 25% protein [10].

From the 1990s onwards, hemp has been established in the food market. The first product was hemp oil, whereas nowadays, there is hemp leaf tea, hemp seeds, hemp oil, beer, lemonade [53], hemp tofu, and hemp milk available [12]. Hemp-fu and hemp milk meet the needs of consumers that seek non-animal protein and milk.

Hempseeds are also used for butter and margarine to substitute expensive fat sources [10]. Dehulled seeds and hemp oil are mainly used for human products [16]. Besides human food products, hemp seeds are also used for animal feed [53].

Whole seeds are the cheapest product of the hemp crop and are mainly used for animal feed [16]. The by-product of hemp seed oils, seed cakes, can also be processed into animal feed [28, 55]. Animals that are fed hemp seed and cakes are birds, fishes [55], ruminants, pigs, horses, poultry, and pigeons. Horses are also fed whole stalks or hemp leaves [16]. Ruminants that were fed hemp seeds and cake as dietary supplements provided milk with a higher fat and energy content [63]. Two other studies also mentioned that hens that are fed hemp seeds lay heavier eggs with a larger egg mass [29, 35].

7.2.3 Paper

Paper can be made of the long bast fibres of hemp stalks [16]. After hemp crops are harvested, the stalks are dried and cleared from the leaves and roots. Once the stalks are dried, they are transported to the pulp mill. In the pulp mill, the stalks are chemically and mechanically processed to extract the fibres for the end product, paper pulp. After this process, the pulp is dried and transported to the paper factory. Industrial hemp has higher environmental impacts than some tree species as the crops are sown annually, which requires both machinery and fertilisers [19]. An advantage of hemp over trees is that the crop can be harvested after four months, whereas a tree requires 20–80 years. Furthermore, chemicals that are used for hemp paper are less toxic than those of wood pulp paper [58].

Hemp paper has some benefits over tree paper, as the paper is stronger, finer [16], and bleached hemp paper will not turn as yellow as conventional paper [58, 73]. Hemp paper is mainly used for speciality paper such as technical filters, banknotes, bible paper, dielectric, and medical paper and cigarette paper [10, 95], but can also be used for office paper [73]. The use of hemp paper on a larger scale is currently hindered by the higher costs of hemp paper compared to conventional paper. According to Crini et al. [16], the high costs of hemp paper can be explained as the bast fibres are only a small percentage of the crop. However, hemp can be grown for dual-purpose seed and fibre cultivation as well. The use of dual-purpose crops often lowers agricultural costs [21].

7.2.4 Textile

Hemp bast fibres are also used for textile products. Compared to other textile fibres, hemp fibres have certain properties which differentiate them from traditional textile fibres. Hemp fibres are stronger, hold their shape more, and stretch less than other natural fibres [66]. The fibres have a higher absorbency and hydrophilicity [50, 65].

High water absorption and hydrophilicity improve the comfort and wear properties, increase fabric durability, and improve the colour strength of hemp fabrics [41]. Textiles made out of hemp fibres have good thermal conductivity [50, 85]. Textiles with high thermal conductivity contribute to the heat transfer of the human body through the textile as the thermal conductivity of air is lower than that of textiles. In other words, the wearer of hemp apparel will stay warm in cold temperatures and stay cool in high temperatures [85]. Furthermore, hemp textiles lack allergenic effects. Consumers that are prone to allergies and with sensitive skins can therefore wear hemp textiles without complications [50]. These consumers should still be careful as chemical dyes and finishes that are added later in the textile process can also cause allergic reactions [38]. Textiles made out of hemp also have good UV protection ability [48, 50]. As ozone depletion is becoming a large problem, the need for textiles with good or excellent UV protection is growing [48].

Two current bottlenecks for hemp fibre and textile production are the need for specialised machines and the relatively high costs of processed hemp fibres [16]. Furthermore, strong fibres can create downsides as well, as the stiff fibres need additional treatments before they can be used for apparel textiles. Improved technology that makes the fibres softer and lighter makes hemp an exciting option for many textile products [66]. Hemp fabrics can be used in the fashion industry for apparel [3, 66], jeans, sportswear [60, 66], underwear, socks, shoes [98], bags, hats [22], accessories, and jewellery [34]. Next to hemp uses for the fashion industry, hemp can also be used as a fabric for furniture, home furnishing [54, 72], lampshades, pillowcases, sheets, duvet covers [52], rugs, and carpets [8].

7.3 Promising Industries

The use of hemp in cosmetic products and the automotive sector is growing, and there is much more potential in these industries [16]. Hemp seed oil contains many minerals, vitamins, and fatty acids, which are beneficial for dry skins [51, 80, 81]. Therefore, hemp is implemented in a broad range of cosmetic products such as make-up, moisturising creams, and hair products [5]. In the automotive industry, hemp is used for car composites, as the need for fibres with a lower environmental impact than glass fibre increased the demand for hemp fibres in the automotive industry [83].

7.3.1 Cosmetics

Hemp oils are made out of hempseed and are rich in minerals, vitamin A, vitamin C, vitamin E, β-carotene, α-linolenic acid, and linoleic acid. Hemp oil is a lightweight ingredient that can be used as a body oil or in lipid creams [51]. The oil of hemp seeds contains high amounts of fatty acids [80]. Therefore, hemp oil can be used as moisturising ingredients in lipid products for dry skins [81]. Another benefit of hemp oil is that it contains natural antioxidants, which reduce the rate of oxidation of vegetable oils and lipid products [11].

Products that are already available on the market that contain hemp oil are facial oils, moisturising face serums, eye creams, priming water, face mists, day creams, face masks, cleansing milk, make-up remover, lotions, body butter, hair oil, hair cream, hair serum, hair styling elixirs, shampoos, conditioners, and hair masks. The brand make-up revolution even created a hemp make-up collection that includes mascara, lip balm, lip-gloss, brow mascara, concealer, foundation, setting spray, and eyeshadow palette [5].

7.3.2 Automotive Sector

In 1941, hemp fibres in combination with other cellulosic fibres were already used as composites for an automobile body of Ford [91]. However, environmental constraints and concurring synthetic resins restricted the future market development for hemp composites. But as soon as the need for more environmentally friendly materials grew, the popularity of hemp composites increased [83]. The advantage of hemp fibres over glass fibres is lower cost, reduced composite weight [39], and the low environmental impacts of hemp as the fibres are renewable and require less energy [69]. Some disadvantages of composites made out of natural fibres such as hemp are the high moisture absorption and low binding to hydrophobic polymer materials [31]. Adhesion of hemp fibres to the polymer materials can be improved through alkaline treatments with sodium hydroxide [87, 88].

Hemp composites can be produced in three ways. In the first method, hemp fibres are mixed with polymer fibres and formed into the desired shape. The second method sprays or soaks nonwoven natural fibre mats with synthetic binders, after which it is pressed into the final form [45]. The last method is through an injection mould, where hemp fibres are mixed with polylactic acid and injected into a mould [70]. Composites made out of hemp are used for certain car models of BMW, Mercedes, Volkswagen, Golf, Peugeot 308, Megane, DS7, Alfa Roméo Giulia [16]. Next to composites for the automotive industry, hemp composites can also be used for furniture and sports goods [72].

7.4 Innovative Technologies

Other industries in which hemp is not as common are in the medical and therapeutic sector, acoustic domain, biofuels, and phytoremediation [16]. Nevertheless, there are many more opportunities for hemp in these industries. Already one medicine that includes CBD has been approved by the FDA [27]. Furthermore, hemp fibres can be used as sound-absorbing materials when implemented in wall panels [40].

Two industries in which hemp is gaining interest due to its environmental performance are the biofuel industry [43, 75] and as a phytoremediation treatment [13, 57].

7.4.1 Medical and Therapeutic Domains

CBD products have many medical and therapeutical benefits for users. CBD is nowadays also used by patients to treat anxiety, insomnia, joint pain and inflammation, arthritis, depression, muscle tension, migraines, chronic pain, and nausea [37]. Clinical evidence exists that the use of cannabidiol may have a positive effect on anxiety [7, 18, 84], arthritis [9], inflammatory pain [92], and nausea [71]. The effect of CBD on insomnia and sleep gives mixed results [89]. Pre-clinical animal studies showed antidepressant effects on mice after the use of CBD [20, 23, 56, 96], but studies on the effect of CBD on human depressions are limited [68]. More controlled clinical trials are needed to prove the effectiveness of CBD [84, 89]. As CBD is becoming more popular, brands are claiming many different benefits and cures of CBD for certain medical conditions without providing evidence [64].

Even though there is currently no evidence of public health-related problems of pure CBD [93], the unfounded claims of CBD products are creating a risk to users as the composition and quality are unknown. Furthermore, uncontrolled products can be contaminated by pesticides and other chemicals that were used during the cultivation, managing, and packaging [64]. Before pharmaceutical drugs are approved, it requires pre-clinical research, clinical research, reviews, and post-market safety monitoring [26, 61]. The US Food and Drug Administration (FDA) warns consumers of the claims and misbranding of these CBD products. There is currently only one CBD medicine that the FDA approved, which is EPIDIOLEX® [27]. EPIDIOLEX® is a medicine that treats seizures related to the Lennox-Gastaut syndrome and Dravet syndrome in 2 years or older patients [33].

7.4.2 Acoustic Domain

Traditional noise control materials are made out of synthetic materials, but the need for more sustainable natural fibres for acoustic purposes is growing [79]. The irregular form and the ellipsoidal or polygonal shapes of hemp fibres make them suitable for sound absorption. Other parameters that determine whether a material is suitable as a noise control material are the porosity and material density [40]. The bast fibres and shives of hemp can be used for music and soundproof panels [72]. Another use of hemp as a soundproof wall is hemp-lime concrete. Hemp-lime concrete showed excellent sound absorption. The porosity and material density have a lower influence on the acoustic properties of hemp-lime concrete than the chemical composition of binders [47].

7.4.3 *Biofuels*

Fossil fuels for heat, power, and vehicles can be replaced by biofuels [36, 77]. The interest in biofuels is growing as the emissions of CO_2 can be mitigated when fossil fuels are replaced by biofuels. The energy input of the crops production chain should have a lower ratio than the output before a crop can be eligible for biofuel. Hemp has good energy output-to-input values and has a higher energy yield per hectare which makes it a suitable crop for biofuel production [43, 75]. Different types of hemp biofuels that can be produced are bioethanol, biodiesel, biogas, and biohydrogen [78].

7.4.4 *Phytoremediation*

Phytoremediation is the treatment of polluted soils with crops and plants, such as hemp [1, 57]. Contaminated soils by heavy chemicals and toxic waste are becoming a larger problem [4, 82, 90, 97]. Characteristics that make hemp a suitable crop for phytoremediation are the high biomass, long roots, short life cycle, and high absorption of heavy metals such as lead, nickel, cadmium, zinc, and chromium [13, 57]. Hemp crops have already been used to clean contaminated soils and extract toxic waste around the Chernobyl plant [32].

7.5 Conclusion

The demand for hemp products is growing as the environmentally friendly crop has a lower impact and higher yield than most of the traditional materials that are used in the building, car, textile, paper, and biofuel industry. Besides these benefits, hemp is also used in the food industry as the seeds of hemp are rich in fat and proteins. Furthermore, demand for dietary supplements will grow as more consumers are looking for healthy or vegan food alternatives. For the cosmetic sector, the minerals, vitamins, and fatty acids from hemp seeds can be beneficial for dry skins. Besides the many benefits of the fibres and seeds, the cannabidiol content of hemp also creates many opportunities in the medical industry as clinical research showed that it could be used to treat anxiety, arthritis, inflammatory pain, and nausea. The following bottlenecks, such as the durability of hemp building materials, the need for specialised machines, and the high costs of processed hemp, will have to be addressed to further stimulate the growth of the hemp industry.

References

1. Ahmad R et al (2015) Phytoremediation potential of hemp (*Cannabis sativa* L.): identification and characterization of heavy metals responsive genes. Clean Soil Air Water 44(2):195–201. https://doi.org/10.1002/clen.201500117
2. Allegret S (2013) The history of hemp. In: Bouloc P, Allegret S, Arnaud L (eds) Hemp—industrial production and uses. CAB International, Wallingford
3. Amaducci S, Gusovius H-J (2010) Hemp—cultivation, extraction and processing. In: Müssig J (ed) Industrial applications of natural fibres: structure, properties and technical applications. Wiley, West Sussex, pp 109–134
4. Arik F, Yaldiz T (2010) Heavy metal determination and pollution of the soil and plants of southeast Tavşanlı (Kütahya, Turkey). Clean: Soil, Air, Water 38(11):1017–1030. https://doi.org/10.1002/clen.201000131
5. Beauty Bay (2021) CBD. Available at: https://www.beautybay.com/l/?q=cbd&f_pg=2. Accessed 27 Mar 2021
6. Bellinger L (1962) Textiles from Gordion. Bull Needle Bobbin Club 46:4–33
7. Bergamaschi MM et al (2011) Cannabidiol reduces the anxiety induced by simulated public speaking in treatment-naïve social phobia patients. Neuropsychopharmacol 36:1219–1226. https://doi.org/10.1038/npp.2011
8. Bhavani K (2015) Performance evaluation of various natural agro fibres in carpet making and their costing. Asian J Home Sci 10(2):296–300. https://doi.org/10.15740/HAS/AJHS/10.2/296-300
9. Blake DR et al (2006) Preliminary assessment of the efficacy, tolerability and safety of a cannabis-based medicine (sativex) in the treatment of pain caused by rheumatoid arthritis. Rheumatology 45(1):50–52. https://doi.org/10.1093/rheumatology/kei183
10. Callaway JC (2004) Hempseed as a nutritional resource: an overview. Euphytica 140:65–72
11. Cantele C et al (2020) Antioxidant effects of hemp (*Cannabis sativa* L.) inflorescence extract in stripped linseed oil. Antioxidants 9(11):1131. https://doi.org/10.3390/antiox9111131
12. Ciano S, Vinci G, Goscinny S (2020) Hemp (*Cannabis sativa* L.): sustainability and challenges for the food sector. Rivista di Studi Sulla Sostenibilita 1:179–194. https://doi.org/10.3280/RISS2020-001010
13. Citterio S, Berta G (2005) The arbuscular mycorrhizal fungus Glomus mosseae induces growth and metal accumulation changes in *Cannabis sativa* L. Chemosphere 59(1):21–29. https://doi.org/10.1016/j.chemosphere.2004.10.009
14. Collet F, Pretot S (2012) Experimental investigation of moisture buffering capacity of sprayed hemp concrete. Constr Build Mater 36:58–65. https://doi.org/10.1016/j.conbuildmat.2012.04.139
15. Costantine G, Maalouf C, Moussa T, Polidori G (2018) Experimental and numerical investigations of thermal performance of a hemp lime external building insulation. Build Environ 131:140–153. https://doi.org/10.1016/j.buildenv.2017.12.037
16. Crini G, Lichtfouse E, Chanet G, Morin-Crini N (2020) Applications of hemp in textiles, paper industry, insulation and building materials, horticulture, animal nutrition, food and beverages, nutraceuticals, cosmetics and hygiene, medicine, agrochemistry, energy production and environment: a review. Environ Chem Lett 18(3):1451–1476. https://doi.org/10.1007/s10311-020-01029-2
17. Crini G, Lichtfouse E, Chanet G, Morin-Crini N (2020) Traditional and new applications of hemp. In: Crini G, Lichtfouse E (eds) Sustain Agric Rev. Springer, Cham, pp 37–88
18. Crippa JAS et al (2011) Neural basis of anxiolytic effects of cannabidiol (CBD) in generalized social anxiety disorder: a preliminary report. J Psychopharmacol 25(1):121–130. https://doi.org/10.1177/0269881110379283
19. da Silva Vieira R, Canaveira P, da Simões A, Domingos T (2010) Industrial hemp or eucalyptus paper? Int J Life Cycle Assess 15:368–375. https://doi.org/10.1007/s11367-010-0152-y

20. de Mello Schier AR et al (2014) Antidepressant-like and anxiolytic-like effects of cannabidiol: a chemical compound of cannabis sativa. CNS Neurol Disord Drug Targets 13(6):953–960. https://doi.org/10.2174/1871527313666140612114838
21. Dube E, Fanadzo M (2013) Maximising yield benefits from dual-purpose cowpea. Food Secur 5:769–779. https://doi.org/10.1007/s12571-013-0307-3
22. Dvorak JE (2004) Hip hemp happenings. J Ind Hemp 9(1):83–88. https://doi.org/10.1300/J237v09n01_09
23. El-Alfy AT et al (2010) Antidepressant-like effect of Δ9-tetrahydrocannabinol and other cannabinoids isolated from *Cannabis sativa* L. Pharmacol Biochem Behav 95(4):434–442. https://doi.org/10.1016/j.pbb.2010.03.004
24. Elfordy S et al (2008) Mechanical and thermal properties of lime and hemp concrete ("hempcrete") manufactured by a projection process. Constr Build Mater 22(10):2116–2123. https://doi.org/10.1016/j.conbuildmat.2007.07.016
25. Evrard A (2008) Transient hygrothermal behaviour of lime-hemp materials. Available at: https://www.researchgate.net/profile/Arnaud-Evrard-2/publication/283568943_Transient_hygrothermal_behavior_of_Lime-Hemp_Materials/links/563f849608ae34e98c4e714c/Transient-hygrothermal-behavior-of-Lime-Hemp-Materials.pdf. Accessed 28 Mar 2021
26. FDA (2019) Development & approval process | drugs. Available at: https://www.fda.gov/drugs/development-approval-process-drugs. Accessed 27 Mar 2021
27. FDA (2020) FDA warns companies illegally selling CBD products. Available at: https://www.fda.gov/news-events/press-announcements/fda-warns-companies-illegally-selling-cbd-products. Accessed 27 Mar 2021
28. Fike J (2016) Industrial hemp: renewed opportunities for an ancient crop. Crit Rev Plant Sci 35(5–6):406–424. https://doi.org/10.1080/07352689.2016.1257842
29. Gakhar N et al (2012) Effect of feeding hemp seed and hemp seed oil on laying hen performance and egg yolk fatty acid content: evidence of their safety and efficacy for laying hen diets. Poult Sci 91(3):701–711. https://doi.org/10.3382/ps.2011-01825
30. Gedik G, Avinc O (2020) Hemp fiber as a sustainable raw material source for textile industry: can we use its potential for more eco-friendly production? In: Muthu SS, Gardetti MA (eds) Sustainability in the textile and apparel industries. Springer, Cham, Basel, pp 87–109
31. Georgopoulos ST et al (2005) Thermoplastic polymers reinforced with fibrous agricultural residues. Polym Degrad Stab 90(2):303–312. https://doi.org/10.1016/j.polymdegradstab.2005.02.020
32. Grauds C, Childers D (2005) The energy prescription. Bantam Dell, New York
33. Greenwich Biosciences (2018) Full prescribing information: contents. Greenwich Biosciences, Carlsbad. Available at: https://www.accessdata.fda.gov/drugsatfda_docs/label/2018/210365lbl.pdf. Accessed 27 Mar 2021
34. Greslehner CAE (2005) Industrial hemp in Austria. J Ind Hemp 10(1):117–122. https://doi.org/10.1300/J237v10n01_10
35. Halle I, Schöne F (2013) Influence of rapeseed cake, linseed cake and hemp seed cake on laying performance of hens and fatty acid composition of egg yolk. J Consum Prot Food Saf 8:185–193. https://doi.org/10.1007/s00003-013-0822-3
36. Havukainen J, Nguyen MT, Väisänen S, Horttanainen M (2018) Life cycle assessment of small-scale combined heat and power plant: environmental impacts of different forest biofuels and replacing district heat produced from natural gas. J Cleaner Prod 172:837–846. https://doi.org/10.1016/j.jclepro.2017.10.241
37. HelloMD and Brightfield Group (2017) Understanding cannabidiol CBD. HelloMD and Brightfield Group, San Francisco. Available at: https://s3-us-west-2.amazonaws.com/hellomd-news/Understanding-Cannabidiol-CBD-Report.pdf. Accessed 28 Mar 2021
38. Heratizadeh A, Geier J, Molin S, Werfel T (2017) Contact sensitization in patients with suspected textile allergy. Data of the information network of departments of dermatology (IVDK) 2007–2014. Contact Dermatitis 77(3):143–150. https://doi.org/10.1111/cod.12760
39. Holbery J, Houston D (2006) Natural-fiber-reinforced polymer composites in automotive applications. JOM 58(11):80–86. https://doi.org/10.1007/s11837-006-0234-2

40. Hui Z, Fan X (2009) Sound absorption properties of hemp fibrous assembly absorbers. Sen'i Gakkaishi 65(7):191–196. https://doi.org/10.2115/fiber.65.191
41. Ibrahim MS, El Salmawi KM, Ibrahim SM (2005) Electron-beam modification of textile fabrics for hydrophilic finishing. Appl Surf Sci 241(3–4):309–320. https://doi.org/10.1016/j.apsusc.2004.07.053
42. Jami T, Karadea SR, Singh LP (2019) A review of the properties of hemp concrete for green building applications. J Cleaner Prod 239:117852. https://doi.org/10.1016/j.jclepro.2019.117852
43. Jasinskas A, Streikus D, Vonžodas T (2020) Fibrous hemp (Felina 32, USO 31, Finola) and fibrous nettle processing and usage of pressed biofuel for energy purposes. Renew Energy 149:11–21. https://doi.org/10.1016/j.renene.2019.12.007
44. Karus M, Vogt D (2004) European hemp industry: cultivation, processing and product lines. Euphytica 140(1):7–12. https://doi.org/10.1007/s10681-004-4810-7
45. Karus M, Kaup M (2002) Natural fibres in the European automotive industry. J Ind Hemp 7(1):119–131. https://doi.org/10.1300/J237v07n01_10
46. Kim SW et al (2019) Meeting global feed protein demand: challenge, opportunity, and strategy. Annu Rev Anim Biosci 7:17.1–17.23. https://doi.org/10.1146/annurev-animal-030117-014838
47. Kinnane O et al (2016) Acoustic absorption of hemp-lime construction. Constr Build Mater 112:674–682. https://doi.org/10.1016/j.conbuildmat.2016.06.106
48. Kocić A et al (2019) UV protection afforded by textile fabrics made of natural and regenerated cellulose fibres. J Cleaner Prod 228:1229–1237. https://doi.org/10.1016/j.jclepro.2019.04.355
49. Kosiński P, Brzyski P, Szewczyk A, Motacki W (2018) Thermal properties of raw hemp fiber as a loose-fill insulation material. J Nat Fibers 15(5):717–730. https://doi.org/10.1080/15440478.2017.1361371
50. Kostic M, Pejic B, Skundric P (2008) Quality of chemically modified hemp fibers. Bioresour Technol 99(1):94–99. https://doi.org/10.1016/j.biortech.2006.11.050
51. Kowalska M, Ziomek M, Zbikowska A (2015) Stability of cosmetic emulsion containing different amount of hemp oil. Int J Cosmet Sci 37(4):408–416. https://doi.org/10.1111/ics.12211
52. La Redoute (2021) Hemp. Available at: https://www.laredoute.co.uk/psrch/psrch.aspx?kwrd=hemp&virtualsite=100#srt=noSorting. Accessed 30 Mar 2021
53. Lachenmeier DW, Walch SG (2005) Analysis and toxicological evaluation of Cannabinoids in hemp food products-a review. Electron J Environ Agric Food Chem 4(1):812–826. https://doi.org/10.5281/zenodo.438133
54. Lamberti DD, Sarkar AK (2017) Hemp fiber for furnishing applications. IOP Conf Ser Mater Sci Eng 254:192009. https://doi.org/10.1088/1757-899X/254/19/192009
55. Leson G (2013) Hemp seeds for nutrition. In: Bouloc P, Allegret S, Arnaud L (eds) Hemp: industrial production and uses. CAB International, Wallingford, pp 229–238
56. Linge R et al (2016) Cannabidiol induces rapid-acting antidepressant-like effects and enhances cortical 5-HT/glutamate neurotransmission: role of 5-HT1A receptors. Neuropharmacology 103:16–26. https://doi.org/10.1016/j.neuropharm.2015.12.017
57. Linger P, Müssig J, Fischer H, Kobert J (2002) Industrial hemp (*Cannabis sativa* L.) growing on heavy metal contaminated soil: fibre quality and phytoremediation potential. Ind Crops Prod 33(1):33–42. https://doi.org/10.1016/S0926-6690(02)00005-5
58. Malachowska E et al (2015) Comparison of papermaking potential of wood and hemp cellulose pulps, vol 91. Annals of Warsaw University of Life Sciences-SGGW. Forestry and Wood Technology, pp 134–137. Available at: http://agro.icm.edu.pl/agro/element/bwmeta1.element.agro-c9eb2861-1d46-4802-9aad-f24e907d5666/c/134_Annals91.pdf. Accessed 7 Mar 2021
59. Marceaua S et al (2017) Influence of accelerated aging on the properties of hemp concretes. Constr Build Mater 139:524–530. https://doi.org/10.1016/j.conbuildmat.2016.11.129
60. McCann J (2015) Environmentally conscious fabric selection in sportswear design. In: Shishoo R (ed) Textiles for sportswear. Woodhead Publishing, Cambrdige, pp 17–52

61. MHRA (2015) Phase I accreditation scheme requirements. Available at: https://assets.publishing.service.gov.uk/government/uploads/system/uploads/attachment_data/file/473579/Phase_I_Accreditation_Scheme.pdf. Accessed 28 Mar 2021
62. Miao C, Hui L-F, Liu Z, Tang X (2014) Evaluation of hemp root bast as a new material for papermaking. BioResources 9(1):132–142. https://doi.org/10.15376/biores.9.1.132-142
63. Mierlita D (2016) Fatty acid profile and health lipid indices in the raw milk of ewes grazing part-time and hemp seed supplementation of lactating ewes. S Afr J Anim Sci 46(3):237–246. https://doi.org/10.4314/sajas.v46i3.3
64. Montoya Z et al (2020) Cannabis contaminants limit pharmacological use of cannabidiol. Front Pharmacol 11(571832). https://doi.org/10.3389/fphar.2020.571832
65. Mustata A, Mustata FSC (2013) Moisture absorption and desorption in flax and hemp fibres and yarns. Fibres Text Eastern Eur 3(99):26–30. Available at: https://yadda.icm.edu.pl/baztech/element/bwmeta1.element.baztech-965248d4-0afe-499e-9e7a-f9686500f5c2/c/FTEE_99_26.pdf. Accessed 30 Mar 2021
66. Muzyczek M (2020) The use of flax and hemp for textile applications. In: Kozlowski RM, Mackiewicz-Talarczyk M (eds) Handbook of natural fibres, vol 2. processing and applications. Taylor & Francis Group, Poznan, pp 147–168
67. Naithani V, Tyagi P, Jameel H, Lucia LA, Pal L (2020) Ecofriendly and innovative processing of hemp hurds fibers for tissue and towel paper. BioRes. 15(1):706–720. https://doi.org/10.15376/biores.15.1.706-720
68. Oberbarnscheidt T, Miller NS (2020) The impact of cannabidiol on psychiatric and medical conditions. J Clin Med Res 12(7):393–403. https://doi.org/10.14740/jocmr4159
69. Panthapulakkal S, Zereshkian A, Sain M (2006) Preparation and characterization of wheat straw fibers for reinforcing application in injection molded thermoplastic composites. Bioresour Technol 97(2):265–272. https://doi.org/10.1016/j.biortech.2005.02.043
70. Pappu A, Pickering KL, Thakur VK (2019) Manufacturing and characterization of sustainable hybrid composites using sisal and hemp fibres as reinforcement of poly (lactic acid) via injection moulding. Ind Crops Prod 137:260–269. https://doi.org/10.1016/j.indcrop.2019.05.040
71. Parker LA, Rock EM, Limebeer CL (2011) Regulation of nausea and vomiting by cannabinoids. Br J Pharmacol 163:1411–1422. https://doi.org/10.1111/j.1476-5381.2010.01176.x
72. Pil L, Bensadoun F, Pariset J, Verpoest I (2016) Why are designers fascinated by flax and hemp fibre composites? Compos Part A Appl Sci Manuf 83:193–205. https://doi.org/10.1016/j.compositesa.2015.11.004
73. Plazonić I et al (2018) Changes in the optical properties of hemp office papers due to accelerated ageing. In: Proceedings of 9th international symposium on graphic engineering and design. GRID Symposium, Novi Sad, pp 121–127. https://doi.org/10.24867/GRID-2018-p14
74. Potin F, Saurel R (2020) Hemp seed as a source of food proteins. In: Crini G, Lichtfouse E (eds) Sustainable agriculture reviews. Springer, Cham, pp 265–294
75. Prade T, Svensson S-E, Mattson JE (2012) Energy balances for biogas and solid biofuel production from industrial hemp. Biomass Bioenergy 40:36–52. https://doi.org/10.1016/j.biombioe.2012.01.045
76. Pretot S, Collet F, Garnier C (2014) Life cycle assessment of a hemp concrete wall: impact of thickness and coating. Build Environ 72:223–231. https://doi.org/10.1016/j.buildenv.2013.11.010
77. Puricelli S et al (2021) A review on biofuels for light-duty vehicles in Europe. Renew Sustain Energy Rev 137:110398. https://doi.org/10.1016/j.rser.2020.110398
78. Rehman MSU et al (2013) Potential of bioenergy production from industrial hemp (*Cannabis sativa*): Pakistan perspective. Renew Sustain Energy Rev 18:154–164. https://doi.org/10.1016/j.rser.2012.10.019

79. Santoni A et al (2019) Improving the sound absorption performance of sustainable thermal insulation materials: natural hemp fibres. Appl Acoust 150:279–289. https://doi.org/10.1016/j.apacoust.2019.02.022
80. Sapino S, Carlotti ME, Peira E, Gallarate M (2005) Hemp-seed and olive oils: their stability against oxidation and use in O/W emulsions. J Cosmet Sci 56(4):227–251. https://doi.org/10.1111/j.1467-2494.2005.00290_2.x
81. Sator P-G, Schmidt JB, Hönigsmann H (2003) Comparison of epidermal hydration and skin surface lipids in healthy individuals and in patients with atopic dermatitis. J Am Acad Dermatol 48(3):352–358. https://doi.org/10.1067/mjd.2003.105
82. Seh-Bardan BJ et al (2013) Biosorption of heavy metals in leachate derived from gold mine tailings using aspergillus fumigatus. Clean: Soil, Air, Water 41(4):356–364. https://doi.org/10.1002/clen.201200140
83. Shahzad A (2011) Hemp fiber and its composites—a review. J Compos Mater 46(8):973–986. https://doi.org/10.1177/0021998311413623
84. Shannon S, Lewis N, Lee H, Hughes S (2019) Cannabidiol in anxiety and sleep: a large case series. Permanente J 23:18–041. https://doi.org/10.7812/TPP/18-041
85. Stanković SB et al (2019) Novel engineering approach to optimization of thermal comfort properties of hemp containing textiles. J Text Inst 110(9):1271–1279. https://doi.org/10.1080/00405000.2018.1557367
86. Stevulova N, Cigasova J, Estokova A, Terpakova E, Geffert A, Kacik F, Singovszka E, Holub M (2014) Properties and characterization of chemically modified hemp hurds. Materials 7:8131–8150. https://doi.org/10.3390/ma7128131
87. Suardana NPG, Piao Y, Lim JK (2011) Mechanical properties of hemp fibers and hemp/pp composites: effects of chemical surface treatment. Mater Phys Mech 11(1):1–8
88. Sullins T, Pillay S, Komus A, Ning H (2017) Hemp fiber reinforced polypropylene composites: the effects of material treatments. Compos Part B Eng 114:15–22. https://doi.org/10.1016/j.compositesb.2017.02.001
89. Suraev AS et al (2020) Cannabinoid therapies in the management of sleep disorders: a systematic review of preclinical and clinical studies. Sleep Med Rev 53:101339. https://doi.org/10.1016/j.smrv.2020.101339
90. Tang J, Chen A (2019) Diagnosis of soil contamination using microbiological indices: a review on heavy metal pollution. J Environ Manage 242:121–130. https://doi.org/10.1016/j.jenvman.2019.04.061
91. Trauffer A (1941) Pinch hitters for defense. Popular Mech 76(6):3
92. Wade DT et al (2003) A preliminary controlled study to determine whether whole-plant cannabis extracts can improve intractable neurogenic symptoms. Clin Rehabil 17:21–29. https://doi.org/10.1191/0269215503cr581oa
93. WHO (2017) Cannabidiol (CBD) pre-review report agenda item 5.2. World Health Organization (WHO), Geneva. Available at: https://www.who.int/medicines/access/controlled-substances/5.2_CBD.pdf. Accessed 28 Mar 2021
94. Wulijarni-Soetjipto N, Subarnas A, Horsten S, Stutterheim N (1999) Cannabis sativa L. In: de Padua L, Bunyapraphatsara N, Lemmens R (eds) Plant resources of South-East Asia: No 12 (1) medicinal and poisonous plants 1. Backhuys Publishers, Leiden, pp 167–175
95. Yao Y et al (2017) Manufacturing technology and application of hemp cigarette paper with dense ash integration. Int Conf Ser Earth Environ Sci 61:012078. https://doi.org/10.1088/1755-1315/61/1/012078
96. Zanelati TV et al (2010) Antidepressant-like effects of cannabidiol in mice: possible involvement of 5-HT1A receptors. Br J Pharmacol 159(1):122–128. https://doi.org/10.1111/j.1476-5381.2009.00521.x
97. Zhang L, Shao H (2013) Heavy metal pollution in sediments from aquatic ecosystems in China. Clean: Soil, Air, Water 41(9):878–882. https://doi.org/10.1002/clen.201200565
98. Zhang H, Zhong Z, Feng L (2016) Advances in the performance and application of hemp fiber. Int J Simul Syst Sci Technol 17(9):18.1–18.5. https://doi.org/10.5013/IJSSST.a.17.09.18

Chapter 8
The Future of Hemp in the Fashion Industry

8.1 Hemp Market

The hemp market can be distinguished between seed, cannabidiol (CBD) and fibre. In 2020, the global industrial hemp market size was estimated to be USD 5.73 billion. The largest part of this market size was dominated by CBD and seed hemp. Hemp seeds and oil are used in, among others, the food and beverage industry and cosmetic industry. The CBD market size was USD 800 million in 2018. Interest in CBD grew in health and wellness products due to the cannabidiols healing properties. North America has the largest market share of CBD with a 49% revenue share [56].

The market size of fibre hemp is smaller than the size of hemp seeds and cannabidiols. The consumer sales of hemp textile in 2018 were USD 1.08 billion. China, the US and Europe are the leading markets for hemp textiles. China dominates the textile market with a share of 78.4%. A large part of these produced textiles in China is used in the country itself for military purposes and clothing production [56].

The hemp market can be analysed on a micro-, meso-, and also on macro-level. The agricultural facilities and processing of hemp fibres are at the micro-level of the industry. Fashion companies and consumers fall under the meso-level. Import and export countries, national legislations, and other governance are part of the macro-level. An overview of these three categories can be seen in Fig. 8.1.

8.2 Micro-Level

The micro-level in the industrial hemp industry is hemp farmers and the textile processing facilities. Agricultural firms can be distinguished between upstream and downstream firms. Upstream firms are mainly farmers and downstream firms are the processing and manufacturing facilities [74].

F. Dhondt and S. S. Muthu, *Hemp and Sustainability*, Sustainable Textiles: Production, Processing, Manufacturing & Chemistry,
https://doi.org/10.1007/978-981-16-3334-8_8

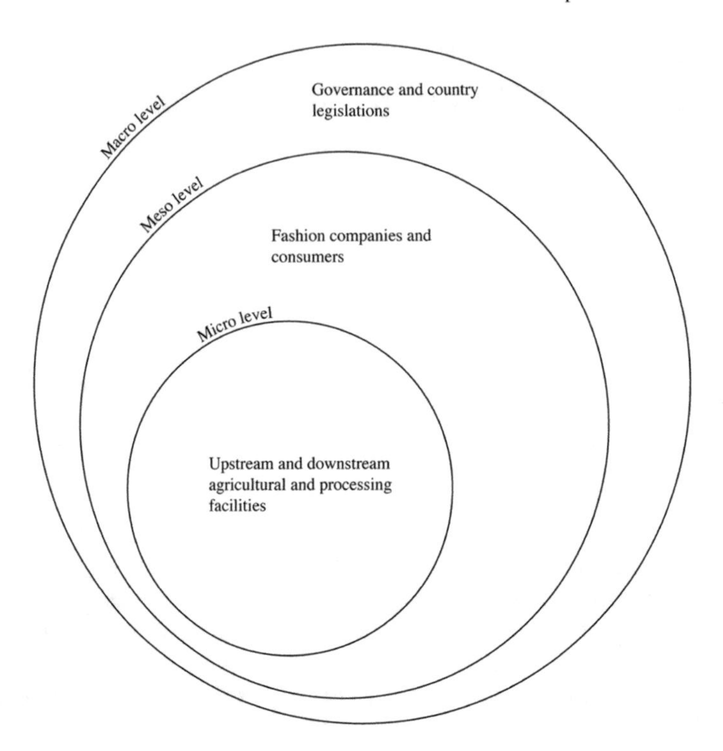

Fig. 8.1 Micro-, meso-, and macro-level of the hemp industry

8.2.1 Downstream

The agricultural market is defined as the perfect competitive model [74]. Hemp farmers offer identical products, hemp fibres, and there are many farmers that sell the same product. High entry barriers are more common for hemp than for other textile fibres, as the crop has been linked to marijuana in the past. Farmers need to obtain licences to grow hemp [36], and the import tariff for hemp fibres is higher than for other natural textile fibres [51–53]. Specialised machines are also required for hemp harvesting as the crop differentiates from other textile crops due to its long and strong stalks [30].

Chinese manufacturers filed for hemp machine patents that reduce labour intensity, waste rate, costs and increase efficiency and possibilities to scale operations. Renewed harvesting equipment is able to separate leaves, seeds, and stalks. The new mechanical machines also recycle hemp seeds that otherwise would be discarded when the crops are harvested for hemp fibres [79]. In the past, harvesting equipment was either labour intensive or only suitable for large-scale farms, as the equipment was very costly. Nowadays, harvesting machines for mid-scale farmers are developed and decrease machinery costs for farmers [35]. Smaller harvest machines are in great demand and create new opportunities for hemp farms [30].

Another innovation that has been created for the hemp market is the blockchain technology trace exchange. Trace exchange allows farmers to keep track of their hemp yield, from soil to shelf and connects farmers to potential buyers. Buyers benefit from the transparent tracking system and the efficient market exchange, which allows them to keep track of their suppliers. Trace was originally developed for the North American market [66] but would be suitable for Europe, Canada, and China as well. Through this platform, suppliers can market hemp products without facing legislation restrictions.

8.2.2 Upstream

Filed patents for industrial hemp innovations peaked from 2019 onwards. Most of the patents for new innovations were filed in China, followed by the US. The largest share of patents is from a subsidiary company of Shanghai Shunhao New Material Technology Co., Ltd 上海顺灏新材料科技股份有限公司 and Luan Yunpeng 栾云鹏 [26, 76]. The hemp fibre processing steps decorticating, drying, and spinning are the largest climate hotspots for hemp yarn and textile production [1, 25, 67, 71, 72, 73, 80].

The first step of processing hemp stalks to fibres is retting [23, 35]. Hemp stalks naturally contain non-cellulosic gums that increase the stiffness of fibres and decrease spinning speed [49]. In the retting process, these non-cellulosic gums are dissolved [35]. Traditional retting methods require a large amount of water and energy, and most retting methods do not meet sustainability requirements [23, 67]. Furthermore, the traditional retting methods are expensive, which impact the costs and price of hemp fibres [23].

New retting technologies entered the hemp market and increased the sustainable potential for retted hemp. More promising innovations that can replace the traditional processes of dew and water retting are seawater retting [81, 82], enzyme retting [77], electrolytic degumming [23] and steam retting [46]. Seawater retting reduces the amount of freshwater that would normally be contaminated with water retting [81, 82]. The most popular eco-friendly retting method, enzyme retting, makes use of microbial enzymes that reduce the non-cellulosic content in the stalks [2, 17, 77]. Electrolytic degumming is another method that is already implemented on a commercial scale in other industries [58]. In this retting method, two components separate hemp fibres through a high-voltage pulse electric discharge (HVPED). Fibres can be pre-treated with biochemicals before being subjected to HVPED [39, 42]. The last method that has been identified as a less polluting retting method is steam retting. Steam retting can be used in combination with an alkaline treatment. The explosion of steam separates the fibres from the stems [22, 46].

Other fibre processing steps, such as the mechanical decortication process, also improve through new innovations. The artificial intelligence-driven system of the Canadian Industrial Hemp Corporation (CIHC) integrates the breaking of stalks together with the processing of other hemp products. The artificial intelligence of the system predicts the potential quality and yield of hemp decorticated bales for bast fibres, hurds, and dust. The end-materials can be selected according to consumer demands by the systems intelligent sensors. The system is expected to lower water and unit costs [16].

Before hemp fibres are spun, an optional process is cottonisation. This process softens the fibres to be able to be spun, knit or woven on regular systems. Modern cottonisation is a combination of retting and ultrasound treatment. This modern method reduces the environmental impact compared to standard cottonisation processes [35]. After the fibres are prepared for spinning, they can be spun on either wet or dry spinning systems. The environmental impact of hemp yarns reduces if the fibres are dry spun, as wet spun yarns are dried after spinning, which requires more energy. Next to that, modern weaving machines will likely reduce energy use and are more efficient than older machines [73].

The last steps of textile processing are dyeing and finishing. Traditional textile dyeing and finishing processes are large contributors to freshwater contamination [37]. Therefore, new dyeing methods with less freshwater contamination are needed to reduce these impacts. A dyeing method that uses 90% less water, 95% less toxic chemicals, and increases production capacity is the dyeing method of ColorZen. This renewed method alters the molecular structure of fibres. The fibres are pre-treated before dying, which increases the bonding of the dye to the fibres and reduces conventional pre-treatment methods that include the use of harsh chemicals. The technology can be used in every dying house and does not require large structural changes or investments. Even though the dyeing method was originally developed for cotton products, it can also be used for other cellulosic fibres [41]. Bleaching of fibres can be done with ozone bleaching, a less energy-intensive method than chlorine bleaching. Ozone bleaching can either be used in gaseous or liquid form. As aqueous ozone bleaching fits better in the current textile processes, it is more commercially viable [38]. Regular bleaching requires warm water, but cool water is already sufficient for ozone bleaching [37]. Hemp fabrics can be softened by bee's wax [65], aloe vera [11], or vitamin E [63] treatments or finishing as a substitute to harsh chemicals [37].

8.3 Meso-Level

Hemp textile is often seen as a sustainable alternative to conventional fabrics for the fashion industry [23]. The fashion industry itself can be divided into fashion companies and consumers. The most valuable fashion companies are Louis Vuitton, Nike, Gucci, Hermès, Zara, Adidas, Chanel, H&M, and Uniqlo [21]. Six out of nine companies have used or use hemp in apparel, footwear, or accessories [14, 24, 27, 31, 47, 62].

8.3.1 Fashion Companies

The current business model of most fashion companies is not sustainable. Companies are mainly focusing on short production times, fast sales, trendy items, and low costs, which all reduce product quality. Environmental pollution due to extensive consumerism and unsustainable business practices intensifies under the current business models. Next to the strategy of individual companies, the industry is under constant cost pressure and high competition. Especially, companies that try to differentiate from other brands by competing on costs struggle with a change to sustainable business manners. The whole industry is responsible for the pollution of the current production processes, and collaboration is needed to reduce the industries environmental impacts [50].

As regulations become stricter and companies become more aware of their environmental footprint, small initiatives to improve the production chain are evolving. The change to a more sustainable industry is at the moment led by fashion producers as they contain the cognitive knowledge and understanding of technical aspects of the supply chain. Depending on the company, designers are often the decision-makers when it comes down to both commercial aspects and the eventual collection of a brand. Designers that are more focusing on trendy products with short lifecycles and business growth will further hold back the change to a sustainable industry. However, designers lack technical knowledge of fibre preparation, textile processing, and sustainable production. An increase in knowledge of designers with regards to these topics will have a large impact on the current business choices as designers are decision-makers for many fashion companies [20].

Reformed business models that focus on limiting growth, reducing pollution and stimulating a circular economy are fundamental methods to change the current industry. The reduction of water, consumption, energy use, chemical use, greenhouse gas emissions, and waste are hindered by the financial investment that is required for these measures. Some environmental measures such as reducing energy and water consumption reduce expenses and are more attractive to business as they benefit from it. Other forms of impact reduction that do not reduce cost and even increase costs, such as sustainable fibres, renewed processing methods, and control technologies, are less attractive to companies. However, these methods are necessary to improve the current industry as well [50].

The second biggest opportunity for fashion companies in 2021 and beyond will be sustainability, according to the State of Fashion Report from Business of Fashion and McKinsey [5]. Clothing will always have an environmental impact, but there are many ways in which the fashion industry can improve its current business practices. The most important parts for brands to focus on are transparency, lowering emissions from upstream operations, reducing the brands own operations, and encouraging sustainable consumer behaviour [9]. A change to materials with lower greenhouse gas emissions, as part of a broader sustainability strategy, was the top agenda point of fashion companies, according to a survey by Berg et al. [8]. An optimal sustainable textile portfolio can only be reached when it contains a diverse fibre range [60]. Hemp would be a great fit for this strategy [23].

Changing the textile use of a company from mainly one fibre to other fibres requires a change in fibre suppliers. Tryouts like pilot programs or capsule collections help brands to make this change [13]. Through capsule collections suppliers, brands and consumers can get familiar with certain products before it is implemented on a large scale [8]. Increased interest in hemp fibres through these pilot programs increases demand for hemp products which will further scale certain innovations [78]. As four out of the five largest fashion companies pledged to change at least half of the materials from their collections to sustainable fibres, the industry shift to the use of sustainable fibres already started on a larger scale [8]. The scale-up of these companies creates opportunities for hemp to be used on a broader scale in the industry.

8.3.2 Consumers

Consumer's behaviour and demand are a large driver for change in the fashion industry. The current consumer behaviour is still focused on extensive consumerism and fast-fading trends [50]. At the same time, a consumer survey by McKinsey reported that 2 out of 3 consumers expect brands to produce sustainable apparel products to mitigate their environmental impact following the corona crisis [7]. Even when fashion brands change their entire textile portfolio, extensive consumerism will degrade the sustainable performance of the product life cycle as garments end up as waste faster. One way to approach consumer demand is by implementing higher prices for eco-friendly products [50]. Currently, the prices for sustainable products are slightly higher than the core products of brands [18].

8.4 Macro-Level

On a macro level, policies of the leading production countries can improve the implementation in the fashion industry. As supporting policies will spike the interest of farms, the supply of the fibre will grow [78].

The leading production countries of hemp are China, South Korea, Russia, the US, and Canada [49]. The tetrahydrocannabinol (THC) content in hemp is regulated in the production countries. The amount of THC that is allowed fluctuates between 0.2% in Europe [48] and 0.3% in the US [70] and China [45]. The level of THC that is allowed in industrial hemp is lower than the average THC value of 5–10% in drug 'genotypes' of the cannabis plant [12].

Data from the Observatory of Economic Complexity OEC [54] shows that France and the Netherlands export most of the hemp fibres, whereas China and Italy are the largest exporters of hemp yarns in 2019. Germany and Austria imported most of the hemp fibres, whereas India and South Korea import most of the hemp yarns in 2019. The total value of hemp fibre imports and export is around USD 33 million, and the imports and exports of hemp yarns are USD 8 million. The fibres import and export value grew from 2018 to 2019 by 26%, and the yarns import and export value increased by 50%. The average import tariff of hemp fibre is 5.68%, and for hemp, yarn is 5.07%. The Bahamas, Romania, Bangladesh, and India have the highest import tariffs for hemp fibres, whereas the highest import tariffs for hemp yarn are in the Bahamas, Sudan, Ethiopia, and Algeria [54].

The import value of Germany and Austria is high due to the automobile industry, which is very active in these countries. Hemp fibres are mainly used there for composite parts of the car [40, 55]. BMW implements hemp composite in the electric i3 and i8 car as it reduces weight and environmental impacts [15]. Furthermore, India and South Korea are the largest importers of hemp yarn as the textile industry is relatively present in these countries [6, 32]. China, the largest cultivator of hemp fibres, is not exporting as many fibres as yarns, as they have well-developed national yarn facilities [6].

The 2018 Farm Bill of the US legalised the production of industrial hemp. The 2018 Farm Bill withdrew the hemp plant from the Controlled Substances Act. Before the 2018 Farm Bill, certain farmers were allowed to grow hemp in pilot programs. From 2018 onwards, the American hemp industry scaled up, and interest grew [43]. In 2019, 200,000 ha were licenced under the new Farm Bill [75]. This amount is expected to be double in 2022 [64]. The US hemp market focuses mainly on CBD hemp, and only 5% of farmers focus on fibre hemp. This percentage is expected to increase with an annual growth rate of 10.5%, as the Biden administration is investing in clean energy and rebuilding infrastructures [78].

In 2020, a trade agreement between the US and China was reached about the import of 216 agriculture products. This trade agreement will further increase the import of China in hemp fibres [69]. As US hemp farmers are currently mainly focusing on CBD hemp, this agreement can increase the interest of farmers in hemp fibres. Currently, farmers face difficulties in finding a market for the left-over hemp biomass after the flowers are cut for CBD [29].

Other government programs that improved the current technologies and innovations of hemp processing were sponsored by Europe and China. The European Union funded the Hemp-SYS and MultiHemp projects that researched bottlenecks of the current hemp industry. The European projects made the registration and development of new genotypes easier, focused on a THC boundary, and designed new, more efficient hemp harvesting equipment's [3, 4]. Chinese Government programs are targeted more on increasing the amount of produced hemp fibres. The eventual goal is to cultivate 1.3 million hectares and yield 2 million tonnes of hemp [81, 82].

8.4.1 Policy Measures

Other policy measures that could further improve the scaling of the hemp industry are clear seed guidance for farmers, increasing the allowed THC content, loosening custom regulations, and changing the current market restrictions.

Improved seed guidance for new farmers will eventually improve hemp yields as farmers will gain valuable knowledge about which genotype is suitable for their preferred end use and climate. Furthermore, the non-regulated selling of hemp seeds resulted in certain farmers that had lower fibre quantity than what was originally promised. Improved regulation will reduce the number of scammers that are active and will increase eventual fibre quality [43].

The second measure that can improve hemp yield is increasing the allowed THC content. The currently allowed value is between 0.2 and 0.3% [45, 48, 70], whereas marijuana drug types contain between 5 and 10% [12]. A slight increase of the allowed THC content will reduce the number of hemp hectares that go 'hot'. The term hot is used to define hemp fields that go above the allowed value of the growing region and which, therefore, will be destroyed [28]. Furthermore, increasing the allowed THC content creates opportunities for new hemp strains for oil and fibre purposes [12]. Increased allowed THC content will also change the main focus of hemp breeding programs from mainly focusing on staying below the 0.2% or 0.3% to improving strains for their suitable end use [48].

The US customs regulations on hemp processing machines are restricting farmers to import harvesting and processing equipment from Europe [29]. European harvesting and processing machines are more developed and advance than the available units in the US [57]. As a result of the import regulations, farmers and innovators are required to develop their own machines, and new innovations are rushing into the market [43, 61]. Despite the inconvenience this import restriction causes to farmers, it can be beneficial for the market as innovators are developing new, more efficient, and productive machines.

Changing marketing regulations for hemp products will further increase consumer awareness of hemp products. Increased awareness about the positive impacts of hemp will reduce the negative stigma that was created when hemp was mistakenly seen as a drug. Certain social media platforms such as Facebook and Instagram are restricting the advertisement of certain hemp and CBD products. The platforms restrict advertisements of drug and drug-related posts that promote the sale or use of illegal, prescription, or recreational drugs [19, 33]. Twitter already changed their policy for Canada and US and allows the advertisement of both CBD and cannabis products under certain regulations [68].

8.5 Future Hemp Market

Scaling of the industrial hemp market will take years, not months. As specialised machines and investment are needed, the establishment of the hemp market will continue in the upcoming years. The demand for sustainable products grows, so investments are expected to rise in the upcoming two years [78]. Increased investments will further allow improvement of current hemp harvesting and processing equipment.

Some of the main trends are in the North American market, China, and the fashion industry. Hemp fibre products with fewer processing steps then hemp textile will first reach market volume in North America; after that, hemp textile products will follow [78]. The Chinese hemp industry is already well established, and the government even sets targets to increase its hemp fibre cultivation further [81, 82]. Furthermore, trends in the fashion industry, between 2018 and 2020, show that the use of hemp in UK and US fashion products grew by 28%, and the year–over-year use of hemp increased by 2% [18]. The three main phases of the future hemp market can be seen in Fig. 8.2.

Phase 1—Investments and improved hemp harvesting and processing equipment
At this time, the hemp production industry is just starting in certain countries and renewed interest in the crop grows. This results in more investments, but farmers and processors are also facing certain bottlenecks. The main bottlenecks that are

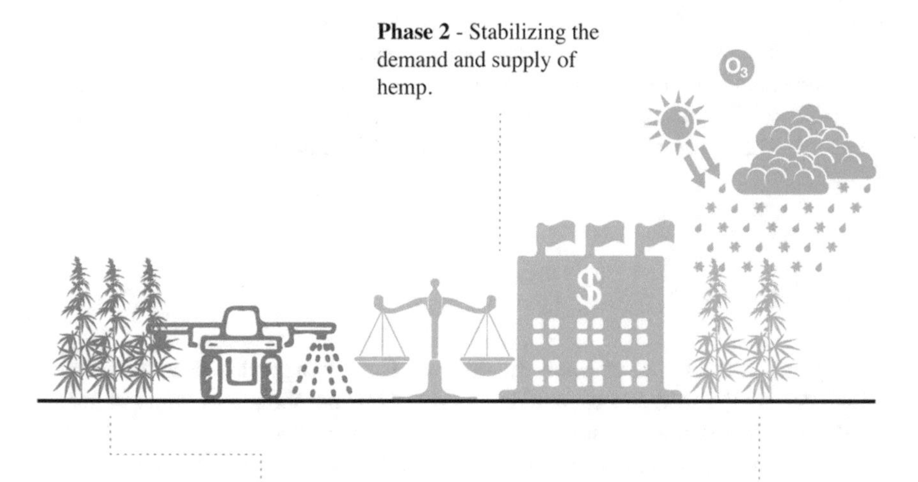

Fig. 8.2 Three phases of hemp's future

holding back scaling of the industry are the outdated and rather costly harvesting and processing units. Therefore, this first phase will focus more on more sustainable and efficient production methods [30]. The hemp market is ready for larger investments [61], and in the next two years, investments in the hemp market are expected to increase [78]. Further development of the hemp market and the renewed interest can increase the uptake of hemp in the textile industry.

Phase 2—Stabilising the demand and supply of hemp
Demand for sustainable products is growing [7], and more apparel products are made of hemp fibres [18], but it will take some years before the market is fully stabilised. The current prospects are set at around 20 years from now [10, 49]. The use of hemp fibres in fashion products is growing over time and will likely grow more in the next couple of years, but before hemp can be at the level of cotton or even linen, it needs to be more economically viable [23]. Increased cultivation and new innovations from the first phase will decrease cost and increase efficiency [61]. Next to that, a new market will form as African production of hemp will likely increase. The focus of this market will be on low to mid-tech applications, such as hemp textiles [44].

Phase 3—Preparing for climate change
The future textile industry will be at risk for climate change. As temperatures will increase by 1.5 °C between 2030 and 2052 [34], it will also be a threat to hemp crops. A shift to sustainable production with less environmental pollution can mitigate worse climate impacts. For the fashion industry, hemp, as a low-input crop, has great benefits over other textile crops [23]. However, farmers should already prepare for new environmental conditions through adaptation methods such as improved hemp strains [59] and more efficient agricultural practices. The fashion and other retail industry should rethink their business models from solely focusing on profits to also prioritise human welfare and environmentally friendly production [23].

8.6 Conclusion

The overall market size of hemp is growing. Even though this growth is mainly stimulated by the seed and cannabidiol market of hemp, the use of hemp in fashion products is increasing. As some of the largest fashion companies stated that they would increase the use of sustainable fibres, the use of hemp textiles might increase even more in the upcoming years. Some current bottlenecks are the rather costly harvesting and processing units and the lack of responsibility that is taken by the fashion industry. Extensive consumerism also holds back the fashion industry to improve its sustainability performance as the main innovations are only focused on increasing operation speed to meet this demand. Nevertheless, the hemp industry is evolving rapidly, and improved efficient technologies, new processing equipment, and government support will further drive the industry forward. Despite the past obstacles, the hemp industry continues to push boundaries and creates machines that will produce hemp textiles faster, cheaper, and with higher quality. The future will likely hold numerous sustainable hemp products and innovations. Now, this is the time for hemp to be implemented.

References

1. Abass E (2005) Life cycle assessment of novel hemp fibre—a review of the green decortication process. Imperial College London, Department of Environmental Science and Technology, London
2. Akin DE et al (2004) Progress in enzyme-retting of flax. J Nat Fibers 1(1):21–47. https://doi.org/10.1300/J395v01n01_03
3. Amaducci S (2003) HEMP SYS: design, development and up-scaling of a sustainable production system for hemp textiles: an integrated quality SYStems approach. J Ind Hemp 8(2):79–83. https://doi.org/10.1300/J237v08n02_06
4. Amaducci S (2017) Multipurpose hemp for industrial bioproducts and biomass. MultiHemp, Milan. Available at: https://cordis.europa.eu/docs/results/311/311849/final1-final-publishable-report-multihemp.docx. Accessed 25 Mar 2021
5. Amed I et al (2020) The state of fashion 2021: Business of Fashion and McKinsey. Available at: https://www.mckinsey.com/~/media/McKinsey/Industries/Retail/Our%20Insights/State%20of%20fashion/2021/The-State-of-Fashion-2021-vF.pdf. Accessed 25 Mar 2021
6. Antczak A, Greta M, Kopeć A, Otto J (2019) Characteristics of the textile industry of two asian powers: China and India. Prospects for their further development on global markets. Fibres Text Eastern Eur 5(137):9–14. https://doi.org/10.5604/01.3001.0013.2895
7. Arici G, Lehmann M (2020) CEO agenda 2020—covid 19 edition. Global Fashion Agenda & McKinsey, Copenhagen. Available at: https://www.mckinsey.com/business-functions/strategy-and-corporate-finance/our-insights/the-restart. Accessed 23 Mar 2021
8. Berg A et al (2018) Fashion's new must-have: sustainable sourcing at scale. McKinsey. Availabe at: https://www.mckinsey.com/~/media/mckinsey/industries/retail/our%20insights/fashions%20new%20must%20have%20sustainable%20sourcing%20at%20scale/fashions-new-must-have-sustainable-sourcing-at-scale-vf.pdf. Accessed 25 Mar 2021
9. Berg A et al (2020) Fashion on climate—how the fashion industry can urgently act to reduce its greenhouse gas emissions. McKinsey & Company and Global Fashion Agenda. Available at: https://www.mckinsey.com/~/media/McKinsey/Industries/Retail/Our%20Insights/Fashion%20on%20climate/Fashion-on-climate-Full-report.pdf. Accessed 26 Mar 2021
10. Boatman-Harrell J (2019) When it comes to sustainability, hemp is the fabric of fashion's future. Available at: https://hypebeast.com/2019/8/hemp-fashion-sustainability-legalization-cotton. Accessed 20 Mar 2021
11. Bolari V (2005) Method for treating textiles and articles of clothing. Italy, Patent No. IT TO20050788 A1
12. Callaway JC (2008) A more reliable evaluation of hemp THC levels is necessary and possible. J Ind Hemp 13(2):117–144. https://doi.org/10.1080/15377880802391142
13. Cernansky R (2020) The impact of fashion's 'sustainable' capsule collections. Available at: https://www.voguebusiness.com/sustainability/impact-fashion-sustainable-capsule-collections. Accessed 21 Mar 2021
14. Chanel (2021) Top cotton & hemp. Available at: https://www.chanel.com/hu/fashion/p/P70821K10109AW001/top-cotton-hemp/. Accessed 25 Mar 2021
15. Colomer-Romero V, Rogiest D, García-Manrique JA, Crespo JE (2020) Comparison of mechanical properties of hemp-fibre biocomposites fabricated with biobased and regular epoxy resins. Materials 13(24):5720. https://doi.org/10.3390/ma13245720

16. Czinner R, Chute W (2019) System, controller, and method for decortication processing. WIPO, Patent No. WO 2021007675 A1
17. Donaghy JA, Levett PN, Haylock RW (1990) Changes in microbial populations during anaerobic flax retting. J Appl Bacteriol 69(5):634–641. https://doi.org/10.1111/j.1365-2672.1990.tb01556.x
18. EDITED (2020) The sustainability EDIT 2020: EDITED. Available at: https://edited.com/wp-content/uploads/2020/11/The-Sustainability-EDIT-2020-Report-2.pdf. Accessed 25 Mar 2021
19. Facebook (2021) 5. Drugs and drug-related products. Available at: https://www.facebook.com/policies/ads/prohibited_content/drugs. Accessed 26 Mar 2021
20. Fletcher K, Grose L (2012) Fashion & sustainability, 1st edn. Laurence King, London
21. Forbes (2020) The 2020 worlds most valuable brands. Available at: https://www.forbes.com/the-worlds-most-valuable-brands/#34d6d3fd119c. Accessed 25 Mar 2021
22. Garcia-Jaldon C, Dupeyre D, Vignon MR (1998) Fibres from semi-retted hemp bundles by steam explosion treatment. Biomass Bioenergy 14(3):251–260. https://doi.org/10.1016/S0961-9534(97)10039-3
23. Gedik G, Avinc O (2020) Hemp fiber as a sustainable raw material source for textile industry: can we use its potential for more eco-friendly production? In: Muthu SS, Gardetti MA (eds) Sustainability in the textile and apparel industries. Springer, Cham, Basel, pp 87–109
24. Goat (2021) Adidas Seeley 'Hemp'. Available at: https://www.goat.com/sneakers/seeley-hemp-g98087. Accessed 25 Mar 2021
25. González-García S, Hospido A, Feijoo G, Moreira MT (2010) Life cycle assessment of raw materials for non-wood pulp mills: hemp and flax. Resour Conserv Recycl 54(11):923–930. https://doi.org/10.1016/j.resconrec.2010.01.011
26. Google Patents (2021) Industrial hemp. Available at: https://patents.google.com/?q=%22industrial+hemp%22&oq=%22industrial+hemp%22&sort=new. Accessed 21 Mar 2021
27. H&M (2021) Twill kostuumpantalon. Available at: https://www2.hm.com/nl_nl/productpage.0930215001.html. Accessed 25 Mar 2021
28. Hamway S (2020) After rocky first year, New Mexico's hemp industry poised to bloom. Available at: https://www.abqjournal.com/1413707/after-rocky-first-year-new-mexicos-hemp-industry-poised-to-bloom.html. Accessed 25 Mar 2021
29. Hemp Benchmarks (2019) U.S. wholesale hemp price benchmarks. Available at: https://cdn-ext.agnet.tamu.edu/wp-content/uploads/2019/09/Hemp-Benchmark-Data-September-2019.pdf. Accessed 25 Mar 2021
30. Hemp Today™ (2019) Developers race to meet demand for hemp-specific technology. Available at: https://hemptoday.net/hemp-specific-technology/. Accessed 20 Mar 2021
31. Hermès (2021) Claudio hat. Available at: https://www.hermes.com/nl/en/product/claudio-hat-H211038NvTR58/. Accessed 25 Mar 2021
32. Hong H, Kang JH (2019) The impact of moral philosophy and moral intensity on purchase behavior toward sustainable textile and apparel products. Fashion Text 6(16). https://doi.org/10.1186/s40691-019-0170-8
33. Instagram (2018) Instagram community guidelines FAQs. Available at: https://about.instagram.com/blog/announcements/instagram-community-guidelines-faqs. Accessed 26 March 2021
34. IPCC (2018) Global warming of 1.5 °C. An IPCC special report on the impacts of global warming of 1.5 °C above pre-industrial levels and related global greenhouse gas emission pathways, in the context of strengthening the global response to the threat of climate change, sustainable development, and efforts to eradicate poverty. World Meteorological Organization, Geneva
35. Jenkins T, Calfee L (2019) Hemp production review of literature with specified scope. Fibershed, San Geronimo. Available at http://fibershed.org/wp-content/uploads/2019/01/hemp-literature-review-Jan2019.pdf. Accessed 28 Mar 2021
36. Johnson R (2014) Hemp as an agricultural commodity. Congressional Research Service, Washington, D.C. Available at: https://apps.dtic.mil/sti/pdfs/ADA599368.pdf. Accessed 25 Mar 2021

37. Kant R (2012) Textile dyeing industry an environmental hazard. J Nat Sci 4(1):22–26. https://doi.org/10.4236/ns.2012.41004
38. Körlü AE (2018) Textile industry and environment. IntechOpen, London
39. Kozlowski R, Baraniecki P, Barriga-Bedoya J (2005) Bast fibres (flax, hemp, jute, ramie, kenaf, abaca). In: Blackburn R (ed) Biodegradable and sustainable fibres. Woodhead Publishing, Sawston, pp 36–88
40. Larson K (2020) The German hemp market—hemp makes a comeback in Germany. United States Department of Agriculture (USDA), Washington, D.C. Available at: https://apps.fas.usda.gov/newgainapi/api/Report/DownloadReportByFileName?fileName=The%20German%20Hemp%20Market%20-%20Hemp%20Makes%20a%20Comeback%20in%20Germany_Berlin_Germany_02-10-2020. Accessed 25 Mar 2021
41. Leonard TM (2013) Treatment of fibers for improved dyeability. WIPO, Patent No. WO 2014116230 A1
42. Maksimov VV (2014) Bast-fiber material processing method. WIPO, Patent No. WO 2016024880 A1
43. Mark T et al (2020) Economic viability of industrial hemp in the United States: a review of state pilot programs. United States Department of Agriculture, Washington, D.C. Available at: https://www.ers.usda.gov/webdocs/publications/95930/eib-217.pdf. Accessed 23 Mar 2021
44. McCann M (2019) The Africa regional hemp & cannabis report—2019 industry outlook. New Frontier Data, Washington, D.C.
45. Mcgrath C (2020) 2019 hemp annual report—Peoples Republic of China. United States Department of Agriculture (USDA), Washington D.C. Available at: https://apps.fas.usda.gov/newgainapi/api/Report/DownloadReportByFileName?fileName=2019%20Hemp%20Annual%20Report_Beijing_China%20-%20Peoples%20Republic%20of_02-21-2020. Accessed 31 Mar 2021
46. Moussa M et al (2020) Toward the cottonization of hemp fibers by steam explosion. Ind Crops Prod, Article in Press, Flame-retardant fibers. https://doi.org/10.1177/0040517517697644
47. MR Porter (2021) Gucci logo-print striped cotton and hemp-blend t-Shirt. Available at: https://www.mrporter.com/en-nl/mens/product/gucci/clothing/striped-t-shirts/logo-print-striped-cotton-and-hemp-blend-t-shirt/9679066509162219. Accessed 25 Mar 2021
48. Müssig J (2010) Industrial applications of natural fibres. Wiley, Bremen. Available at: http://priede.bf.lu.lv/grozs/AuguFiziologijas/Augu_resursu_biologija/gramatas/Industrial%20Applications%20of%20Natural%20Fibres.pdf. Accessed 23 Mar 2021
49. Muzyczek M (2020) The use of flax and hemp for textile applications. In: Kozlowski RM, Mackiewicz-Talarczyk M (eds) Handbook of natural fibres, volume 2: processing and applications. Taylor & Francis Group, Poznan, pp 147–168
50. Niinimäki K et al (2020) The environmental price of fast fashion. Nat Rev Earth Environ 1(1):189–200. https://doi.org/10.1038/s43017-020-0039-9
51. OEC (2021a) Flax fibers. Available at: https://oec.world/en/profile/hs92/flax-fibers. Accessed 25 Mar 2021
52. OEC (2021b) Hemp fibres. Available at: https://oec.world/en/profile/hs92/hemp-fibers. Accessed 25 Mar 2021
53. OEC (2021c) Prepared cotton. Available at: https://oec.world/en/profile/hs92/prepared-cotton. Accessed 25 Mar 2021
54. OEC (2021d) Trend explorer. Available at: https://oec.world/en/visualize/line/hs92/export/show/all/11530820/2013.2019/. Accessed 19 Mar 2021
55. OEC (2021e) What does Austria export? (2019). Available at: https://oec.world/en/visualize/tree_map/hs92/export/aut/all/show/2019/. Accessed 26 Mar 2021
56. Ooyen C (2019) The global state of hemp: 2019 industry outlook. New Frontier Data, Washington, D.C.
57. Pari L, Baraniecki P, Kaniewski R, Scarfone A (2015) Harvesting strategies of bast fiber crops in Europe and in China. Ind Crops Prod 68(1):90–96. https://doi.org/10.1016/j.indcrop.2014.09.010

58. Rutledge N (2018) Engineering chemistry. ED—Tech Press, Essex
59. Salentijn EMJ et al (2015) New developments in fiber hemp (*Cannabis sativa* L.) breeding. Ind Crops Prod 68:32–41. https://doi.org/10.1016/j.indcrop.2014.08.011
60. Sandin G, Roos S, Johansson M (2019) Environmental impact of textile fibers—what we know and what we don't know. Mistra Future Fashion, Stockholm. Available at: http://mistrafuturefashion.com/wp-content/uploads/2019/03/Sandin-D2.12.1-Fiber-Bibel-Part-2_Mistra-Future-Fashion-Report-2019.03.pdf. Accessed 25 Mar 2021
61. Savills Research (2020) Hemp cultivation in the UK. Savills Research, London. Available at: https://www.savills.co.uk/landing-pages/landscope/HempSpotlight.pdf. Accessed 25 Mar 2021
62. Size? (2021) Nike air zoom type 'hemp'—size? Exclusive. Available at: https://tinyurl.com/yjcsmb76. Accessed 25 Mar 2021
63. Son K, Yoo D, Shin Y (2014) Fixation of vitamin E microcapsules on dyed cotton fabrics. Chem Eng J 239(1):284–289. https://doi.org/10.1016/j.cej.2013.11.034
64. Steenstra E (2019) The U.S. hemp market 2019 states ranking. New Frontier Data, Washington D.C.
65. Szulc J et al (2020) Beeswax-modified textiles: method of preparation and assessment of antimicrobial properties. Polymers 12(2):344. https://doi.org/10.3390/polym12020344
66. Trace (2020) Our manifesto. Available at: https://tracevt.com/manifesto/. Accessed 19 Mar 2021
67. Turunen L, van der Werf HMG (2006) Life cycle analysis of hemp textile yarn. Comparison of three fibre processing scenarios and a flax scenario. INFRA, UMR SAS, Rennes
68. Twitter (2021) Drugs and drug paraphernalia. Available at: https://business.twitter.com/en/help/ads-policies/ads-content-policies/drugs-and-drug-paraphernalia.html. Accessed 26 Mar 2021
69. U.S. Trade Representative (2020) Economic and trade agreement between the government of the United States of America and the government of the People's Republic of China. U.S. Trade Representative and Chinese Ministry of Agriculture and Rural Affairs, Washington, D.C. Available at: https://ustr.gov/sites/default/files/files/agreements/phase%20one%20agreement/Economic_And_Trade_Agreement_Between_The_United_States_And_C. Accessed 31 Mar 2021
70. USDA (2020) Hemp and farm programs. Available at: https://www.farmers.gov/sites/default/files/documents/USDA_OptionsforHemp-Factsheet-02062020.pdf. Accessed 20 Mar 2021
71. van der Werf HMG, Turunen L (2008) The environmental impacts of the production of hemp and flax textile yarn. Ind Crops Prod 27(1):1–10. https://doi.org/10.1016/j.indcrop.2007.05.003
72. van der Werf HMG (2004) Life cycle analysis of field production of fibre hemp, the effect of production practices on environmental impacts. Euphytica 140(1):13–23. https://doi.org/10.1007/s10681-004-4750-2
73. van Eynde H (2015) Comparative life cycle assessment of hemp and cotton fibres used in Chinese textile manufacturing. KU Leuven, Leuven
74. Vettas N (2006) Market control and competition issues along the commodity value chain. In: Governance, coordination and distribution along commodity value chains. Food and Agriculture Organization (FAO), Rome, pp 9–26. Available at: http://www.fao.org/3/a1171e/a1171e.pdf. Accessed 20 Mar 2021
75. Vote Hemp (2020) 2019 U.S. hemp license report. Vote Hemp, Washington, D.C. Available at: https://www.votehemp.com/wp-content/uploads/2019/09/Vote-Hemp-US-License-Report-2019.pdf. Accessed 26 Mar 2021
76. WIPO (2021) Patentscope. Available at: https://patentscope.wipo.int/search/en/result.jsf?_vid=P12-KMJKOO-33932. Accessed 21 Mar 2021
77. Yadav D et al (2016) Potential of microbial enzymes in retting of natural fibers: a review. Curr Biochem Eng 3(2):89–99. https://doi.org/10.2174/221271190302160607151925
78. Yahn-Grode T, Morrissey K, McCann M (2021) The U.S. hemp market landscape—cannabinoids, grain & fiber. New Frontier Data, Washington D.C.

79. Yufeng S, Li S, Ge Y (2018) Industrial hemp combine harvester. China, Patent No. CN108243712
80. Zampori L, Dotelli G, Vernelli V (2013) Life cycle assessment of hemp cultivation and use of hemp-based thermal insulator materials in buildings. Environ Sci Technol 47(13):7413–7420. https://doi.org/10.1021/es401326a
81. Zhang J (2008) Natural fibres in China. In: Common fund for commodities proceedings of the symposium on natural fibres. Food and Agriculture Organization (FAO) and Common Fund for Commodities (CFC), Rome, Italy, pp 53–61. Available at: http://www.fao.org/3/i0709e/i0709e.pdf. Accessed 25 Mar 2021
82. Zhang LL et al (2008) Seawater-retting treatment of hemp and characterization of bacterial strains involved in the retting process. Process Biochem 43(11):1195–1201. https://doi.org/10.1016/j.procbio.2008.06.019

Printed in the United States
by Baker & Taylor Publisher Services